Hydrogen Fuel Cells

This book focuses on the deployment of hydrogen fuel cell technologies to ensure efficient and reliable performance during cold start conditions. It encompasses an introduction to hydrogen energy systems and the unique challenges associated with cold starts in fuel cells. It examines catalytic heaters, electrolyte-heating circuits, pre-heating of reactant gases, electric resistance heating, self-heating through load cycling, and hydrogen burners. Each chapter delves into the principles, design, and performance of these systems, highlighting their advantages, limitations, and applications. The chapters of this book:

- elaborate on specific challenges and practical solutions for improving fuel cell reliability and efficiency in low-temperature environments;
- provide a comparative analysis of cold start solutions;
- discuss the latest advancements in materials, design techniques, and industrial applications;
- explore topics including catalytic heaters, pre-heating systems, and self-heating mechanisms;
- include practical applications and case studies.

This book is aimed at graduate students and researchers in chemical engineering, hydrogen fuel processing, and energy integration.

Hydrogen Fuel Cells

Cold Start Solutions

Vamsi Krishna Kudapa
Surajit Mondal
S. Siddharth

CRC Press
Taylor & Francis Group
Boca Raton London New York

CRC Press is an imprint of the
Taylor & Francis Group, an **informa** business

Designed cover image: Getty

First edition published 2026
by CRC Press
2385 NW Executive Center Drive, Suite 320, Boca Raton FL 33431

and by CRC Press
4 Park Square, Milton Park, Abingdon, Oxon, OX14 4RN

CRC Press is an imprint of Taylor & Francis Group, LLC

ISBN: 978-1-041-10627-2 (hbk)
ISBN: 978-1-041-11014-9 (pbk)
ISBN: 978-1-003-65786-6 (ebk)

DOI: 10.1201/9781003657866

Typeset in Times
by SPi Technologies India Pvt Ltd (Straive)

Contents

About the Authors

Vamsi Krishna Kudapa, Associate Professor, has been working in the Department of Chemical and Petroleum Engineering, University of Petroleum and Energy Studies (UPES), Dehradun, Uttarakhand, India, since 2013. Before UPES, he worked as an Assistant Professor and Head of the Department of Petroleum Engineering, Sri Aditya Engineering College, Andhra Pradesh, India. He pursued his Ph.D. in Oil & Gas – Modelling and Simulation from UPES, Dehradun, in 2018. In his 10 years of experience, labs such as the Drilling Fluids & Cementation Lab and the Petroleum Product Testing Lab were developed. In addition, a unique prototype model, "Digital Oil Field," was developed at UPES in 2015, which is a combination of gas lift and oil-spill control mechanism apparatus. So far, he has 40 publications, 5 books, 2 patents, and 10 book chapters to his credit. He is presently working on hydrogen energy, solar energy applications in agriculture, nanoparticles applications in drilling fluids and cementation, enhanced oil recovery, modeling, and simulation of unconventional gas reservoirs.

Surajit Mondal completed his Master's in Energy Systems and PhD in the Renewable Energy domain in 2020. Currently, he is teaching at the UPES to UG, PG, and PhD Scholars. He has published more than 60 international research and review articles as an author/co-author. He has published 65 patents and has been granted 38 patents in his name. He has completed two DST (Government of India) funded projects of approximately INR 1 crore (or 10 million) in the field of Energy Systems/Sustainability.

S. Siddharth is an educationalist and researcher working in R&D projects relating to mechanical engineering. Born in Tiruchirappalli, India, he has completed his bachelor's in engineering from SASTRA University, and PG and Doctorate from Anna University, Chennai, India. He has published 72 books, 169 papers in refereed national and international journals. His areas of interest include advanced optimization, application of machine learning, and artificial intelligence in manufacturing processes.

Preface

In an age characterized by the need to deal with problems such as global warming mitigation, long-term sustainability, and energy security, hydrogen is among the most promising energy carriers that can transform the global power landscape. The combination of its natural abundance, its high gravimetric energy density, and the unique property of giving only water as a byproduct from combustion or electrochemical conversion positions hydrogen as the key to a transition to renewable and sustainable energy systems worldwide. A critical yet underappreciated facet of Hydrogen Fuel Cells' Cold Start Solutions is explored: Fuel cells can work reliably and efficiently under cold start conditions. This capability is not just worth technical curiosity; it is a key parameter to the commercial viability, safety, and scalability of hydrogen-fed applications, for transport, distributed power generation, and prime industrial sectors.

Hydrogen fuel cells are considered a key technology toward deep decarbonization that could replace fossil-fuel-based systems in many applications. However, their wide adoption at scale has been hampered by performance issues under extreme and poor environmental conditions. Of these, low-temperature operation, especially initial start-up of the fuel cell in cold climates, remains among the longest-running technical bottlenecks. Water that forms as a reaction byproduct can freeze during cold starts in the internal parts of the fuel cell and cause efficiency loss, long warm-up times, membrane and catalyst layer degradation, and mechanical damage in extreme cases. All such issues may degrade the short-term performance and long-term durability and become a big hindrance in facing the market competition. Here, science, engineering, and practical solutions to these barriers are explored. Its scope is intentionally broad: it will be a reference for researchers at the frontier of fuel cell innovation, engineers working in system integration, students entering the field, and industry professionals looking for commercial approaches.

The opening chapters review the fundamental thermodynamic and electrochemical challenges limiting fuel cell performance at sub-zero temperatures and place these challenges in the context of hydrogen as a clean and sustainable energy vector. From this background, the text explores cold start solutions such as catalytic heaters for fast thermal ramp-up; electrolyte-heating circuits for ice prevention in membrane electrode assembly; pre-heating reactant gases for optimum reaction kinetics; electric resistance heating for localized and controlled thermal input; self-heating via load cycling to exploit internal heat generation during start-up; and hydrogen burners for auxiliary thermal sources for fast de-icing/system stabilization. Each approach is reviewed from fundamental principles through material selection, through design trade-offs, through integration strategies to operational safety, while considering the most recent advances in materials science, such as low-temperature ionomers, nanostructured catalysts, and thermally conductive composite materials. In addition to technical solutions, the book deals with cross-cutting issues of next-generation hydrogen fuel cell systems, such as energy efficiency and parasitic power consumption minimization during cold start, sustainability through responsible material sourcing and life-cycle optimization, hydrogen handling safety protocols for cold environments, and scalability/manufacturability in mass-market applications.

To address the break between theory and practice, real-life case studies of the automotive, stationary, and aerospace divisions are presented as invaluable experience in both the implementation issues and the performance results. These examples represent not only the technological possibility but also the economic and regulatory implications of adoption. Focusing on intense scientific review and drawing upon experience through sound engineering advice, *Hydrogen Fuel Cells: Cold Start Solutions* hopes to go beyond being a technical booklet; it is to be an agent of change. With global hydrogen demand ramping up as the world seeks to meet decarbonization objectives, the capacity to start and operate fuel cells reliably regardless of climate will play an important role in the ability to fully harness the potential of hydrogen.

Are you an established practitioner interested in tweaking the performance of operating systems, a scholar venturing into the uncharted territory of hydrogen technology, or a level-headed consumer eager to see what all the fuss in clean energy is about? This book welcomes you to immerse yourself in one of the most thrilling and influential tasks in the discipline. We hope this knowledge may not only educate but also motivate you to act on the collaborative projects required to create a cleaner, greener, and more resilient energy future powered by hydrogen.

1 Cold Start Assisted Systems (Catalytic Heaters)

1.1 INTRODUCTION TO COLD START CHALLENGES IN FUEL CELLS

The start-up of hydrogen fuel cells is a major operational problem. This affects the electrochemical performance and system readiness, mainly in freezing environments. In low-temperature domains, starting hydrogen fuel cells remains a challenge, and the sub-zero climate adds to the starting difficulties (Ertekin et al., 2024). Unlike conventional power mechanisms, hydrogen fuel cells need exact water and thermal regulation. Therefore, electrochemical reactions can start and continue successfully. Below zero degrees, water changes its intracellular behavior dramatically. This phase change affects fuel cell function. These ice blockages are critical to structural elements like microporous layers and internal flow channels. This obstruction prevents the transmittance of reactant gases to the catalyst surface. This interference decreases active reaction surface area, and mass transport disruption is exacerbated (Guan et al., 2023). Meanwhile, pressure distribution is distorted inside the cell. The gas stoichiometry is also changed. These further delay system ignition. Meanwhile, proton conductivity of the polymer electrolyte membrane decreases under cold stress, which slows ion transport but increases internal cell resistance. This degradation leads to a voltage and power drop in the early operational phase, and the system's efficiency is affected. It becomes more complicated when onboard heating mechanisms work very slowly. The resulting thermal lag further extends the delay to functional equilibrium. Therefore, in practice, this situation delays system response time and decreases the drivability of a fuel cell electric vehicle (FCEV). This causes component stress as well as uneven heating/abrupt thermal changes. To get around these operational barriers, cold start technologies must be applied strategically, and their combination increases reliability. Catalytic heaters provide a promising heating approach based on controlled hydrogen amounts for localized thermal energy generation by catalytic oxidation. This technique heats important cell elements efficiently. This targeted warming mechanism is proven effective in energy-sensitive applications. The system complexity is preserved while the fuel usage is low (Pramuanjaroenkij & Kakaç, 2023). Advancement of fuel cell stack architecture, improved insulation materials, and smart algorithmic controls reduce the negative effects of low-temperature environments. A good cold start framework should consider thermal as well as material behavior. It's necessary to ensure safe and reliable activation of hydrogen fuel cells under freezing operation constraints.

DOI: 10.1201/9781003657866-1

1.1.1 Impact of Low Temperatures on Fuel Cell Performance

Low-temperature operation of hydrogen fuel cells poses several reliability and performance challenges. Those challenges were accentuated in Proton Exchange Membrane Fuel Cells (PEMFCs). The electrochemical and physical properties of internal components change significantly under sub-zero thermal conditions. The changes occurring within the system impacts system efficiency and structural integrity. A critical problem is the freezing of operational water in regions such as flow channels, gas diffusion layers, and membrane electrode assemblies (MEAs), where it turns mechanically into ice and causes stress on the membrane and catalyst interfaces, leading to layer delamination and microcracks. This mechanical damage damages fuel cell materials permanently. This worsening affects the performance and operational life of the fuel cell stack (Tao et al., 2025). Also, fundamental electrochemical reactions like hydrogen oxidation or oxygen reduction remain very sensitive to temperature. The reactions of these decrease markedly in colder environments. Meanwhile, proton conductivity decreases within the polymer membrane as ionic mobility is limited. This creates more resistance between the cell structural layers. Such parallel conditions induce voltage reductions and lower electrical output in early operational phases. This considerably limits the system response. Uneven or mismanaged thermal input further degrades the problem. Variation in thermal expansion rate increases the membrane rupture possibility. Rising resistance plus slowed reactions combine to make adequate power in cold climates difficult to obtain. This limitation becomes important in fuel cell vehicles, which require a fast and reliable power supply. For these cold start problems, systems must include effective purging strategies to remove residual water prior to shutdown. This prevents ice from developing. Water repellent materials and frost-resistant catalysts were widely used, while fast, even heating methods like catalytic heaters greatly improved cold start performance. Meanwhile, intelligent flow-field architectures and optimized thermal management schemes preserved fuel cell operation during and after freezing temperatures (Zhang et al., 2022).

1.1.2 Start-Up Time and Efficiency Losses

Starting time and overall system efficiency are two key performance metrics that directly affect the efficiency of hydrogen fuel cells for transport and stationary applications. These two parameters also denote their commercial success. The cold start condition introduces operational delays in variable climate environments. These delays prevent the fast realization of stable functionality. During such transitional phases, the system balances stack temperature, gas flow consistency, and membrane hydration level. The entire start-up process requires a significant amount of energy input because auxiliary components, such as pumps, heaters, and compressors, must operate before the fuel cell can produce usable electricity. Its duration depends on the thermal inertial system, ambient temperature, and preconditioning strategies. All these affect the operational readiness. Since different hydrogen fuel cell technologies have their own operational mechanisms and since start-up timelines vary widely across applications, this leads to application-specific delays. High-temperature cell types, such as solid oxide and molten carbonate fuel cells, require extended warm-up times to reach the high operational temperatures necessary for good ionic conduction and

strong reaction kinetics. Conversely, low-temperature alternatives such as PEMFCs allow faster initiation but require strict moisture regulation to avoid membrane flooding or dehydration. Such conditions degrade the performance when not controlled. Solutions are pre-heated by catalytic or resistive thermal methods to speed up start-up. Such techniques save energy (Ferrara & Hametner, 2021). Efficiency degradation influences initial system activation and long-term steady-state function. This decline results from different energy loss pathways. Much external energy is needed to warm the fuel cell stack at start-up. This external demand decreases the net electrical yield. In normal operation, auxiliary energy consumption, partial hydrogen utilization, and systemic heat losses still affect efficiency negatively, and these limitations are enduring. Material degradation, like catalyst sintering, membrane fatigue, and contaminant buildup, led to performance degradation with ageing effects, which require frequent monitoring. Innovations like high-efficiency thermal exchangers, dynamic control algorithms, and advanced thermal systems may mitigate such inefficiencies. They are adaptive and therefore enable real-time operation tuning. Lags during start-up, as well as ongoing efficiency losses, must be reduced for the commercial viability of hydrogen fuel cell technologies.

1.1.3 Icing and Water Management Issues

The behavior of water in an electrochemical system influences the performance and operational lifetime of hydrogen fuel cells, especially those using proton exchange membranes (PEMs). This influence increases with environmental conditions. A good water regulation is therefore very important during cold start events. It will also appear as a product of electrochemical reactions and as an important conduit for proton transport. Once the operating temperature is below freezing, the remaining water from prior cycles or new water formed during start-up turns into ice and blocks important internal components such as flow channels, catalyst surfaces, and microporous diffusion layers. This freezing behavior restricts reactant mobility and raises mass transport resistance significantly. This blocks electrochemical reaction activity. Consequently, the voltage output of the system is reduced substantially, and ice accumulation leads to mechanical stress due to repetitive freeze–thaw cycling. These mechanical stresses deteriorate the MEA (Qu et al., 2024). This damage worsens fuel cell stability. Even amidst freezing conditions, ineffective water regulation leads to operational problems in medium to high-current load scenarios. Water accumulation due to wrong drainage or inconsistent gas delivery floods porous regions, and this flooding prevents oxygen access to the surface of the cathode catalytic surface. Such interference compromises the triple-phase boundary critical to a good reaction efficiency and reduces the electrochemical result. In the opposite direction, inadequate water availability expels the membrane material and raises internal resistance due to restricted proton conduction under dry or low-humidity conditions. The balance between membrane hydration and water evacuation remains important to prolong system functionality and maintain uniform current distribution. Systems employ advanced dynamic management strategies to mitigate these operational water risks. Those strategies shall include innovations like hydrodynamically optimized flow fields, intelligent humidification mechanisms, and gradient-porosity diffusion layers for real-time adaptation. Sensors and thermal imaging are also used for diagnostic purposes, such as ice development tracking and internal water distribution pattern observation. Material-level advancements like hydrophilic protective coating and membranes for water retention improve resilience under various environmental conditions. This makes hydrogen fuel cells more adaptable, which makes them suitable for a wide climate conditions and practical use cases.

1.1.4 Need for Auxiliary Heating Systems

Hydrogen fuel cells, particularly those based on PEM technology, show serious performance limitations at low temperatures, and this operational constraint makes the addition of supplementary heating mandatory. These thermal systems are more than just supporting parts, but are essential for a reliable start-up and consistent performance under cold and hostile ambient conditions. PEM fuel cells operate in the narrow thermal range of 60°C to 80°C and have good ionic conductivity and excellent electrochemical kinetics in this band. Below this specified temperature range, performance is poor, and membrane hydration is a problem, which reduces system effectiveness. This makes auxiliary heating support indispensable for the fuel cell stack to reach the optimal operating temperature uniformly and rapidly.

Early start-up phases are generally in freezing environments and involve internal ice buildup risks and thermal shocks. This threatens cell integrity. Water retained from prior operations or formed during system initiation freezes and expands. This expansion creates tremendous mechanical pressure. That kind of pressure causes MEA structural distortion or delamination. These effects impair long-term durability. Supplementary heating mechanisms correct this problem by gradually heating important components up before the fuel cell is fully functional. This stabilizes the internal humidity distribution greatly. These conditions greatly reduce the risk of stress-related system failures (McKinlay et al., 2021).

In mobile systems requiring responsiveness and operation dependability, such heaters minimize start-up delays and energy waste due to prolonged warm-up intervals. They are useful to maintain temperature uniformity when waste heat recovery is inconsistent or poorly optimized. Besides traditional resistive heating methods, nowadays, smart thermal storage media, catalytic pre-heaters, and sensor-integrated temperature controls are added to improve system behavior. Such innovations improve thermal management efficiency as well as durability by reducing destructive freeze–thaw frequency. Auxiliary heating subsystems are the critical thermal bridge between shifting environmental conditions and controlled thermal specifications required for stable fuel cell performance.

1.2 PRINCIPLES OF CATALYTIC HEATING

Catalytic heating of hydrogen fuel cell systems was based on the principle of hydrogen oxidation enhancement without traditional direct combustion mechanisms. It was thus an energy-efficient solution for thermal management. In this technique, the special properties of catalysts – mainly noble metals like platinum or palladium – meant that hydrogen could react with oxygen at relatively lower temperatures. Hydrogen is adsorbed on the catalyst surface as it enters the catalysis heating chamber. This weakening of molecular bonds weakens them and causes dissociation to atomic hydrogen. The atomic hydrogen combines with oxygen molecules adsorbed on the catalyst surface to give water vapor. The whole process releases a lot of thermal energy, and energy output is stably controlled via surface-mediated pathways (Liu et al., 2022). As opposed to open flames and fast oxidation methods, catalytic heating produces no visible flame and thus reduces localized overheating, fire hazards, and explosion risks. This safety advantage makes the catalytic heaters suitable for enclosed/compact fuel cell environments. Uniform thermal regulation is mandatory to protect sensitive components and increase system durability. Flameless catalytic heating produces low emissions of airborne particulates and nitrogen oxides. This outcome links catalytic heating practices to global sustainability targets for clean and efficient energy deployment.

The unique efficiency of catalytic heating lies in its precise targeting of heating zones, and this localized delivery accelerates warm-up in cold start scenarios. Thus, the thermal energy resulting from this is directed to critical sections like humidification units, end plates, or coolant conduits, and this spatial targeting helps the system to reach operational temperature very quickly; and the whole heating sequence is adjusted by regulated hydrogen flow rates. This characteristic enables real-time thermal modulation based on embedded sensor data. Catalytic heating as part of the overall thermal management framework allows a reliable start-up in low-temperature environments and improves waste heat utilization. This, in turn, contributes to fuel cell stack conditioning. Currently, it is the central element in advanced hydrogen fuel cell systems.

1.2.1 Catalytic Reaction Mechanism with Hydrogen

Cold start problems in hydrogen fuel cell systems demand catalytic reactions of H_2. It becomes necessary when incorporated in auxiliary heating units. Unlike conventional stacks that must initiate internal electrochemical reactions, catalytic heaters use surface-based H_2 oxidation and operate independently of electrical circuits. This happens as the hydrogen gas flows into a catalyst chamber coated with palladium or platinum and spreads out over a porous ceramic substrate.

The catalyst surface provides sites on which hydrogen molecules can adsorb and break apart into atoms of atomic hydrogen that react with oxygen adsorbed onto the surface. Oxygen comes from ambient air or a regulated supply. The reaction is a Langmuir–Hinshelwood mechanism with the two reactants adsorbed onto the catalytic surface. It is a strongly exothermic surface-level reaction that produces water vapor and constant thermal energy. This involves no open flames – heat is spread uniformly at moderate temperatures – below 250°C. It is a controlled heat source, which is useful for cold start conditions and warms the fuel cell end plates and the MEA. Total surface exposure, catalyst distribution, and resistance to contaminants such as carbon monoxide and some sulfur compounds are important factors affecting catalytic oxidation performances and reliability (Zion et al., 2021). The output of thermal energy can be adjusted by controlling oxygen and hydrogen flow rates, enabling system-level thermal management. In other words, catalytic oxidation in hydrogen heaters is an indirect but effective thermochemical technique suitable for hydrogen fuel cells instead of pure electrochemical heating systems. It improves operational readiness and provides thermal support where needed.

1.2.2 Role of Catalyst Materials (e.g., Platinum Group Metals)

Platinum group metals (PGMs) are an important component of hydrogen fuel cell technologies and exhibit excellent catalytic properties at low temperatures, as observed in PEMFCs. They are of catalytic importance during cold start operations and are very useful to start oxygen reduction and hydrogen oxidation reactions with little heat input. Metals like rhodium [Rh], platinum [Pt], and palladium [Pd] were effective and showed good adsorption and excellent surface activity for oxygen and hydrogen molecules, respectively. All this reduces the activation energy required for electrochemical reactions, and platinum is very resistant to acidic media. This quality allows stable performance in repeated freeze/freeze cycles and durability in PEMFC systems in various cold conditions. In catalytic heaters, PGMs drive exothermic hydrogen–oxygen reactions and produce heat that preconditions the fuel cell stack. A platinum catalyst supported on ceramic substrates controls reaction rates and distributes heat evenly without introducing an open flame. Here, the catalyst's function is extended to include thermal stability and operation safety at low temperatures, besides increasing reaction speed. Also, high material cost and limited natural availability of these rare metals prevent large-scale adoption.

Poisoning, sintering, or catalyst dissolution performance losses affect long-term durability and are usually related to thermal fluctuation exposure and moisture (Wang et al., 2022). Against this background, further research on alloy-based catalysts like Pt-Ni and Pt-Co will be conducted, which will maintain high catalytic activity without compromising total PGM usage. At the same time, efforts to non-precious substitutes like metal–organic frameworks and transition metal oxides are cost-effective and environmentally sustainable. Catalyst architecture, fabrication techniques, and support material development still require further progress toward the commercial viability of hydrogen fuel cell technologies (Table 1.1).

TABLE 1.1
Comparisons of PGM Catalysts in Fuel Cell and Catalytic Heater Applications

Catalyst Material	Primary Use in Fuel Cell	Thermal Stability	Catalytic Efficiency	Cost	Cold Start Suitability
Platinum (Pt)	Anode and Cathode (HOR, ORR)	High	Excellent	Very High	Excellent
Palladium (Pd)	Anode Catalysts and Gas Purifiers	Moderate	Good	High	Good
Rhodium (Rh)	Specialized Heat-Resistant Roles	Very High	Moderate	Very High	Moderate
Pt-Co Alloy	Cathode (Enhanced ORR)	High	Excellent	High	Excellent
Non-PGM (e.g., Fe-N-C)	Emerging Alternative Catalysts	Moderate	Developing	Low	Under Evaluation

1.2.3 Exothermic Reaction Characteristics

A very exothermic reaction is possible at the heart of hydrogen fuel cell functionality. This reaction is needed to make heat and electricity very efficiently. This spontaneous process begins when oxygen meets hydrogen, which lets out a large amount of usable thermodynamically favorable energy. Unlike traditional combustion engines, which produce mechanical energy by thermal expansion, hydrogen fuel cells produce chemical energy directly as electricity via electrochemical means.

This mechanism reduces energy losses of the multi-stage conversion system and improves the system efficiency. In this system, hydrogen gas enters the anode compartment and is oxidized. This splits hydrogen molecules into protons and electrons. That reaction releases energy. The resulting protons and electrons travel across the PEM and create an electric current in an external electrical circuit. An oxygen molecule is reduced at the cathode, and the resulting water is formed with the returning protons and electrons. This redox mechanism involves oxidation at the anode and reduction at the cathode, achieving both thermal energy plus electricity. Thus, the heat produced contributes to the system's thermal profile, especially during cold start conditions when internal components need to be warmed up. Further improvements to this process are introduced using platinum-based catalysts, which reduce the activation energy needed for chemical reactions. They help to accelerate the reaction rates even at low temperatures and are very important to system performance. This catalytic effect is favorable for the hydrogen-powered catalytic heaters, and it is more about thermal generation than about electrical output. The whole process is clean and nondestructive and produces only water as its byproduct. This makes hydrogen fuel cells sustainable and safe within operational protocol limits (Malik et al., 2023).

1.2.4 Advantages over Electrical or Combustion Heating

Hydrogen fuel cells have an advantage in heating applications – in catalytic cold start systems. They beat combustion-based heating systems and traditional electric resistance heaters. Unlike conventional systems that generate heat using resistive elements/open flames, fuel cells perform double duty. Their combined action produces heat and electricity by an electrochemical process. Combined output increases system efficiency and is especially useful in cogeneration or Combined Heat and Power (CHP) systems, where waste heat is recovered and reused, instead of being wasted. Electric heating devices are efficient at the point of use; however, they use electricity from the grid, and this electricity is produced from fossil fuel sources. This kind of upstream energy production introduces conversion inefficiencies and raises carbon emissions, which reduces the sustainability of what looks like a clean energy source. On the other hand, hydrogen fuel cells create no sulfur oxides, carbon dioxide, particulate pollutants, or nitrogen oxides and are therefore suitable for applications in which air quality is vital.

This property makes hydrogen fuel cells suitable for use in restricted spaces, fuel cell–powered cars, residential buildings, and portable energy systems, where clean air is desired with minimal harmful emissions. Catalytic heating uses surface oxidation of H_2 in combination with catalysts for safety and efficiency advantages. It creates heat without flames. This avoids fire risks and decreases the risk of thermal instabilities or sudden temperature changes (Nguyen & Shabani, 2021). Another benefit is the almost silent operation of fuel cells, compared with the noise of combustion engines/gas burners. This silent operation provides user comfort and allows installation in noise-sensitive urban environments/other quiet zones. In addition to this, hydrogen-based heating allows decentralized deployment and thus allows for low-infrastructure off-grid and emergency backup systems. While issues concerning hydrogen storage and transport remain, overall performance benefits of catalytic and fuel cell heating solutions position them as alternatives to combustion-based heating technologies and conventional electricity.

1.3 DESIGN AND COMPONENTS OF CATALYTIC HEATERS

Catalytic heaters embedded in hydrogen fuel cell systems provide steady, flameless heat under various environmental conditions and are very efficient during cold start phases. These heaters regulate hydrogen oxidation with a catalyst and produce thermal energy without using flames. At its core is a reactor module containing the catalytic substrate for hydrogen–oxygen interactions.

Porous or honeycomb structures are used for maximizing system efficiency; such geometries increase available surface area, gas mixing, and heat distribution uniformly over the reaction zone. The catalytic surface is usually infused or coated with noble metals like palladium or platinum. Moreover, because these materials are expensive, research efforts have been directed to more cost-effective and impurity-resistant materials like perovskite compounds or cerium oxide (Luo et al., 2022).

Within this catalytic core, heat recovery, gas flow, and system reliability controls are incorporated. A critical element is the flow distribution manifold, which spreads hydrogen and air over the catalytic surface. This distribution avoids under-reacted zones or hot spots, causing structural damage/energy inefficiency. Many manifold geometry optimizations are carried out for laminar flow with uniform pressure drop in the reactor bed using computational fluid dynamics (CFD). Thermal energy from the catalytic reaction does not get wasted. Instead, a gas heat recuperator collects it and recycles it to warm incoming hydrogen and air. This recycling increases thermal efficiency and saves energy for system start-up. Thermal insulation reduces heat loss and protects surrounding components from elevated temperatures. Temperature sensors and automatic valves that adjust reactant flow rates to maintain the catalyst temperature for safe and responsive thermal output are included. During the material selection, engineers consider corrosion resistance, thermal conductivity, and mechanical toughness. Modern heater designs emphasize modularity and easy integration with existing fuel cell stacks for better adaptability and simple application in various applications.

1.3.1 Catalyst Bed and Substrate Design

Catalytic heaters designed for cold start application in hydrogen fuel cells require a precise combination of components/design strategies and are engineered to provide heat efficiently and safely at low temperatures. They are based on thermochemical processes that produce heat by oxidation of very small quantities, in contrast to electrochemical catalyst layers embedded into MEAs of these heaters. In this system, a catalyst substrate is present, typically a metallic mesh or high-surface-area ceramics coated with catalytic materials such as palladium, platinum, or their alloys. Often, this surface geometry is expressed as wash-coated monoliths, microporous foams, or sintered mesh layers designed to improve the contact of catalytic surfaces with reactant gases.

Proper gas flow distribution is critical for heater operation, and the inlet system delivers ambient air or hydrogen uniformly across the catalyst to avoid partial reactions or uneven heating. For this kind of precision, designers often perform CFD simulations involving flow paths, pressure drops, and turbulence evaluation. Also, well-positioned baffles and integrated flow guides maintain constant residence times and stable reaction environments. Heat generated during operation is controlled by advanced insulating materials and thermal barriers to reduce heat loss and protect the neighboring fuel cell components. Temperature sensors or thermocouples for monitoring real-time thermal conditions are fitted in the heater and send data to control units controlling ignition sequences and gas flow rates. Recently, miniature control electronics were designed directly in the heater module to allow instantaneous hydrogen dosing adjustments based on preset temperature thresholds. The heater housings are made of corrosion-resistant and light metals like aluminum base alloys or stainless steel for long thermal cycles. All of them always allow quick start-up, precise thermal regulation, and safe integration in hydrogen fuel cell systems (Rossetti & Tripodi, 2024) (Table 1.2).

TABLE 1.2
Design Variants in Catalytic Heater Systems for Hydrogen Fuel Cells

Design Element	Variant Type	Function	Advantages
Catalyst Substrate	Ceramic Monolith/ Sintered Metal Mesh	Supports catalyst coating; maximizes contact area	High surface area, thermal stability
Catalyst Material	Pt/Pd/Pt-Pd Alloy/ Non-PGM	Facilitates hydrogen oxidation	High activity, tailored selectivity
Gas Flow Design	Laminar/Turbulent/ CFD-Optimized Channels	Controls reactant distribution	Prevents hot spots, ensures complete reaction
Insulation and Encasing	Aerogels/Fiberglass/ Metal Sheath	Minimizes heat loss	Compact form, enhances thermal efficiency
Temperature Monitoring	Thermocouples/ Infrared Sensors	Provides real-time thermal feedback	Enables precise heating control
Control Integration	Onboard PID/External ECU Interface	Modulates gas flow and heating rate	Adaptable to fuel cell system architecture
System Form Factor	Modular Cartridge/Flat Plate/Cylindrical	Adapts heater geometry to space constraints	Scalability, easy replacement, or retrofitting
Safety Mechanisms	Pressure Relief Valve/ Hydrogen Shutoff	Prevents overpressure or unintended ignition	Enhances system safety and reliability

1.3.2 Thermal Insulation and Housing Materials

For hydrogen fuel cell systems with catalytic heaters for cold starts, suitable housing materials and appropriate thermal insulation are mandatory, and each of these components serves two purposes: They retain heat produced by catalytic hydrogen oxidation to accelerate fuel cell stack warming, and they reduce heat loss to the environment, an important reduction of heat loss during cold climate start-ups.

In accordance with their positioning around heaters or in hydrogen storage tanks or stack assemblies, insulation can operate at elevated and cryogenic temperature conditions. For cryogenic applications like liquid hydrogen storage, multilayer insulation (MLI) with spacer layers and alternating reflective foils in vacuum-sealed structures is generally used, reducing conductive and radiative heat transfer significantly. This design prevents boil-off gas formation, causing pressure buildup and hydrogen loss as well (Ali et al., 2024).

Insulation materials for catalytic heater units must resist high temperature and thermal uniformity without limiting hydrogen flow or reaction efficiency. Suitable materials for these applications are ceramic fiber blankets/aerogels and microporous silica with low thermal conductivity and good chemical stability in a hydrogen-rich atmosphere. They are light and thus suitable for portable/mobile fuel cell systems where weight reduction is desired. Meanwhile, the sensitive components and housings of heaters are mechanically tough, can withstand thermal cycling, and resist hydrogen-induced embrittlement (Tang et al., 2021). For heater and stack housings, austenitic stainless steel, a nonmagnetic, corrosion-resistant material, is used. In high-temperature systems such as Solid Oxide Fuel Cells (SOFCs), advanced materials like engineered ceramics and metal–ceramic composites are applied for structural integrity at temperatures above 800°C. Insulation and housing material development remain fundamental priorities as hydrogen fuel cells move toward higher efficiency and commercialization, and these enhancements are necessary for precise thermal regulation and safe operation in a harsh environment. Figure 1.1 shows how insulation and housing materials serve complementary but different functions, with each having its own role to play in reducing boil-off gas and ensuring electrical isolation for safe system operation.

1.3.3 Gas Flow Management Systems

The gas flow management systems are always in charge of distributing, delivering, and using reactant gases like air and hydrogen throughout the fuel cell stack. These systems control the gas stoichiometry to achieve electrochemical reactions at the cathode as well as at the anode. They reduce waste and prevent the degradation of cell components. For this purpose, some key elements are coordinated together, namely, high-accuracy sensors, mass flow controllers, and solenoid valves. They all operate with a special intelligent control unit. They all control flow rates and pressures and respond to changing power demands. Then, they respond to external environmental conditions like temperature or altitude variations.

The other important task of gas flow control is the water management within the system, and poorly balanced flow results in improper membrane hydration – either

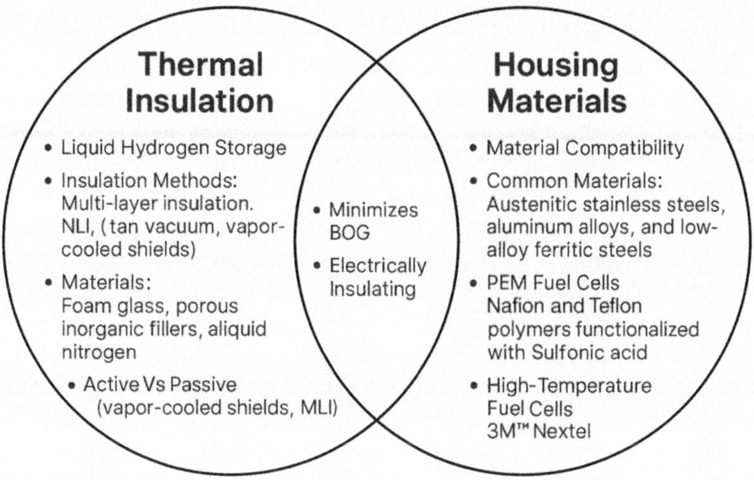

FIGURE 1.1 Venn diagram showing roles and properties of thermal insulation and housing materials in hydrogen fuel cell systems.

by flooding gas diffusion layers with excess water or by drying the polymer electrolyte. Both extremes present challenges. Ionic conductivity is reduced by dry membranes, while excess water prevents oxygen access to catalysts. For overcoming such problems, gas flow systems are often combined with water and humidity management components such as water separators or humidifiers, which work together to maintain the optimum hydration and to remove the accumulated liquid water. Safety is a prime concern in system design due to the high diffusivity and very low ignition threshold of hydrogen. In this respect, automatic shut-off mechanisms, hydrogen leak detectors, and overpressure relief valves are added to mitigate the hazards. A central control unit monitors all pressure, flow rate, and temperature parameters continuously. Those changes are in real time for system security and efficiency. Whether for FCEV powering, stationary energy systems, or portable energy units, a well-designed gas flow management setup plays an extremely important role in performance, safe fuel cell operation, system life expectancy, and reliability guarantee in all usage cases (Maiket et al., 2022).

1.3.4 Compactness and Integration Constraints

Catalytic heater integration in hydrogen fuel cell systems presents a set of design and space challenges, with all of them needing proper management to ensure safe and effective usage across applications. Though the actual hydrogen fuel cells themselves are configurable and modular according to the desired power requirements, the other subsystems, namely, temperature control and gas flow, often place major layout and space constraints on the layout configurations. The main obstacle is hydrogen storage, which requires high-pressure tanks at 700 bar and a cryogenic tank at −253°C. For both approaches, bulky insulated storage solutions are required, which make integration within compact platforms such as unmanned drones, passenger cars, and aircraft difficult.

The catalytic heaters must be space-efficient when combined. They are positioned so that heat transfer to the fuel cell stack is not restricted by electrical wiring, air circulation, or other thermal components. Though flameless and compact in design, catalytic heaters require supporting infrastructure like insulation layers, gas supply lines, flow regulators, and safety detection systems. All these make up the system footprint. Also, other necessary subsystems like humidifiers, power electronics, control units, and pressure regulators must be incorporated well, usually within restricted spatial confines. Also, a compact but functional layout requires creative design strategies and component size reduction (Meda et al., 2023). Further complexity stems from the materials used: high reactivity of hydrogen necessitates corrosion-resistant materials that are more expensive and bulkier than conventional materials. Optimal membrane hydration without adding system bulk and water management during operation will remain a challenge. To overcome these limitations, researchers applied component integration wherever possible and invented multifunctional materials. Innovations like integrated gas flow manifolds, printed circuit heat exchange, and adaptive thermal interface materials are successful. Despite efforts toward smaller hydrogen systems, spatial and integration challenges must be solved for advancing hydrogen-powered technologies toward viability and wider adoption.

1.4 HYDROGEN REGULATION AND CONTROL MECHANISMS

Any good hydrogen fuel cell system must have good hydrogen regulation and control. They guarantee energy efficiency, operational safety, and long-term reliability. Such systems control how hydrogen is delivered, circulated, and consumed while preserving pressure, temperature, and electrical output within acceptable limits. This regulation process starts at the storage interface, where hydrogen is prepared for usage either as compressed gas or cryogenic liquid. Storage pressures vary between 350 and 700 bar, depending on the system design. Before entering the fuel cell, hydrogen is depressurized and measured with mass flow controllers and pressure regulators. These components ensure that hydrogen reaches the anode at the correct pressure and flow rate with respect to changing power demands.

For fuel efficiency and system performance reasons, unreacted hydrogen is often recirculated by passive ejectors or mechanical pumps. Most ejectors work on pressure differentials instead of moving parts and are thus less complex and have lower energy losses. Yet, ejectors always need precise control strategies to keep the flow constant. By balancing hydrogen inputs with oxidant inputs, the stoichiometric ratio is controlled to avoid fuel starvation that damages the membrane and leads to significant performance reduction.

Heat management is another important aspect of hydrogen regulation, since the electrochemical process produces excess heat that should be emitted through cooling loops as well as heat exchangers to stay away from overheating and/or membrane dehydration. A central Fuel Cell Control Unit (FCCU) coordinates the whole operation, synchronizing several subsystems like humidification, hydrogen supply, oxidant flow, and power conversion modules. The FCCU uses real-time sensor data like humidity, temperature, pressure, and flow rates with feedback-driven control logic for stable operation conditions. In the latest fuel cell vehicles, energy management systems employ hybrid strategies combining batteries with fuel cells for energy efficiency. With the help of adaptive control and machine learning, new systems can provide predictive diagnostics and smarter regulation, allowing fuel cells to adapt to the changing loads and environmental variables.

1.4.1 Hydrogen Flow Meters and Valves

In hydrogen fuel cell systems, hydrogen flow meters and control valves are mandatory for system efficiency, fuel delivery accuracy, and safe operation. These parts cooperate to monitor and control the flow of hydrogen gas entering the fuel cell stack, where electrochemical conversion changes it into electricity. Flow meters provide live data on hydrogen flow rate and volume, quick load adjustments, and accurate energy balancing. Typical types are Coriolis mass flow meters, which are resistant to flow interference and accurate; thermal mass flow meters, which measure flow by heat transfer; and ultrasonic flow meters, which read flow with sound waves for non-intrusive readings. Each meter type has its own advantages depending on flow range, pressure, temperature, and system design.

Complementing the flow meters are hydrogen control valves, which mechanically start, stop, or fluctuate hydrogen flow according to system requirements or programmed algorithms. They are pressure-resistant and hydrogen embrittlement-resistant valves that are robust in high-pressure and low-temperature ranges. For cryogenic liquid hydrogen systems, specialized valves with vacuum-insulated housings and low thermal conductivity materials are used. For gaseous hydrogen systems, flow management roles are performed by proportional, solenoid, and check valves. Particularly, accurate valve control is required during transient operations such as cold starts, where improper dosing leads to membrane dehydration or catalytic flooding (Yu et al., 2023). In the hydrogen delivery system, the core is made of sensors, flow meters, and control valves controlled by a central control unit. This system processes sensor inputs for setting hydrogen flow rates and H_2/O_2 ratios for fuel cell operation. Flow meters and valves together improved the responsiveness, the longevity, and the reliability of hydrogen fuel cell systems, mainly in dynamic environmental conditions and variable loads.

1.4.2 Sensors and Feedback Control Loops

In hydrogen fuel cell systems, especially in cold start scenarios, sensors and feedback control loops are mandatory for temperature, reactivity, and system performance management. In such systems, a network of calibrated sensors continuously monitors oxygen and hydrogen concentrations, temperature distribution, humidity levels, gas pressures, and power outputs for various operational parameters like oxygen and hydrogen concentrations. A centralized control unit collects this sensor information in real time and responds accordingly via compressors, valves, actuators, heaters, and other control elements. This ongoing adjustment keeps the system balanced and avoids gas flooding, fuel starvation, and thermal instability.

Hydrogen detection sensors, usually based on metal oxide semiconductors or catalytic pellets, are placed along the hydrogen supply line and in external enclosures to detect even the smallest leaks and stop fire hazards. Their short response time and high selectivity make them suitable for hydrogen detection in confined or cold environments. In parallel, oxygen sensors near the cathode air inlet adjust airflow for electrochemical reactions to maintain a safe and effective hydrogen/oxygen ratio. Thermal sensors in fuel cell stack/coolant loops monitor temperature variations. This is especially important when the catalytic heaters work under low-temperature start-up conditions to achieve localized thermal regulation. Central to system controllers with feedback loops are those controlled by Proportional-Integral-Derivative (PID) algorithms based on PID algorithms. Such loops consistently compare sensor readings against target values and adjust system behavior accordingly. In more advanced configurations involving external disturbances/nonlinear dynamics, advanced control strategies like fuzzy logic/Model Predictive Control (MPC) are applied. Additionally, feedforward control mechanisms are being incorporated to predict load variation/adjust reactant flow. In combination with each other, sensor-driven feedback systems extend system lifetime and reliability even under extreme conditions or challenging environmental stresses to ensure fuel cells always work within design specifications (Hossain et al., 2023).

1.4.3 Start and Stop Triggering Logic

In hydrogen fuel cell systems, especially those with catalytic heaters supporting cold starts, start/stop logic is very important for energy efficiency, system readiness, and the health of the fuel cell stack. Such logic handles idle/active state transitions based on defined parameters such as environmental conditions, power demand, and system diagnostics. The sequence should be timed during start-up, especially in cold environments. A catalyst needs to pre-heat the fuel cell stack to functional temperature while controlling the flow of hydrogen and oxidant gases to prevent membrane dehydration or ice accumulation. Usually, the heater is activated in advance on ambient temperature readings/algorithmic forecasts based on general usage behaviors like daily driving habits. This ensures the stack reaches the thermal threshold before electrochemical reactions begin.

The main control framework, usually contained within the fuel cell electronic control unit (FCECU), processes data from many sensors. It monitors real-time values like voltage output, hydrogen pressure, gas flow rate, and temperature to decide the start/stop of the system. Start-up logic is activated by the ignition key, throttle, low battery levels, or pre-operation diagnostic events. In reverse, the stop sequence is activated during inactive periods when system anomalies are detected during regenerative braking phases, such as gas leaks, electrical fluctuations, or unsafe temperature rises. All these transitions should be quick and intelligent to minimize mechanical strain and energy waste. Advanced control strategies use predictive algorithms to analyze future behavior patterns and driving loads. Various control techniques refine these transitions. For example, fast on/off cycling is prevented by hysteresis-based switching. Advanced approaches use real-time models for predicting system dynamics and fine-tuning inputs: MPC. Start/stop logic will, in many systems, be integrated with hybrid architectures. That means the battery and fuel cell share energy. This cooperative management allows easy transitions between power delivery, fuel efficiency, and protection of critical components, making the system robust and effective for mobile and stationary hydrogen applications.

1.4.4 Safety Shutoff and Leak Detection Systems

In hydrogen fuel cell systems, safety shutdown mechanisms and leak detection protocols are important. It's because the intrinsically risky properties of hydrogen are high diffusivity, a broad flammability range, and a very low ignition threshold. These traits require active and vigilant systems for the detection and prevention of potential fire, explosion leaks in crowded or enclosed areas. Modern hydrogen fuel cell technologies use robust safety architectures combining smart decision-making controls, continuous monitoring, and fast-acting mechanical safeguards. Most important safety features include an automatic valve that cuts off hydrogen flow within milliseconds upon sensing of a system integrity breach or fault. These valves are electronically controlled and depend on a constant flow of feedback from a group of control units that receive data from sensors measuring pressure variations, temperature anomalies, hydrogen levels, and irregular flow patterns.

They are programmed with elaborate safety algorithms and have fail-safe modes as well as redundant circuits to remain functional in case of system failure or damage to the power system. Other designs involve dual valve systems or localized isolation mechanisms, where a leak to one compartment is contained without stopping the whole fuel cell system. To detect leaks reliably in various environmental conditions, the detection network uses different sensor technologies. Semiconductor hydrogen sensors detect leaks by changing electrical resistance. The catalytic bead sensors track exothermic reactions and perform reliably in temperature variations. Parallel thermal conductivity sensors detect minor changes in gas composition. An ultrasonic detector extends coverage by listening for escaping gas high-frequency noise signatures. These leak detection systems are integrated with onboard diagnostics of mobile platforms such as FCEVs and send real-time alerts to drivers via sound alarms or a dash indicator. For stationary and industrial use, these systems are interconnected with building safety infrastructure and provide immediate access to exhaust fans, visual and audible alerts, and maintenance teams. Respecting international safety standards like ISO 26142/IEC 60079 is critical, and experts recommend routine sensor calibration/periodic inspections and layered safety strategies. The strength and reliability of such safety shutdown and leak detection systems are ultimately critical to public trust and compliance with regulatory expectations, making hydrogen a secure, scalable energy solution.

1.5 INTEGRATION WITH FUEL CELL STACK AND SUBSYSTEMS

Integrating catalytic heaters in the fuel cell stack and its subsystems is a challenging engineering problem during hydrogen fuel cell cold start operations. They produce flameless heat by catalytically oxidizing hydrogen on their surface. They are incorporated in the system layout for efficiency. Therefore, the aim is to provide enough thermal energy for cold starts without compromising system safety/compactness/efficiency. Typically, integration starts with placing the catalytic heater near the fuel cell stack. It is generally placed near or along end plates where the generated heat spreads out uniformly across the electrochemical cells. This thermal support is necessary to raise the membrane and catalyst layer temperature above freezing to allow electrochemical reactions to proceed under sub-zero start conditions. This likewise transfers heat efficiently, avoiding localized overheating. A thermally coupled heater is attached to the stack via indirect heating loops or conductive interfaces. Integration with a broader thermal management system involves coordination with elements such as coolant pumps, radiators, and heat exchangers, which allow distribution and balancing of heat input from the catalytic unit with the thermal demands of the fuel cell at operation. Parallel to this, hydrogen and air supply systems match heater fuel requirements for precise delivery during cold starts without interfering with normal fuel cell operation or undesirable gas crossings. This balance is often obtained using sophisticated flow regulators and purging mechanisms.

The control electronics manage this integration and use real-time sensor feedback, including pressure, temperature, and flow-based readings, to control the heater's activity and start-up safely. Transient instability or voltage dips are prevented by synchronization with stack control logic and power electronics. In hybrid configurations, the system must adjust battery energy management strategies to match heater load requirements. Combined with good integration performance, catalytic heaters improve cold start reliability, improve long-term system durability, and reduce the energy penalty of start-up events, which will contribute to wider practical use and adoption of hydrogen fuel cells in real-world conditions (Sultan et al., 2023).

1.5.1 Placement of Heaters in the Stack Environment

The location of catalytic heaters in the fuel cell stack environment influences the energy consumption, cold start efficiency, and the life expectancy of key components, and in hydrogen fuel cell systems such as PEMFCs, uniform and fast heating of the MEA under freezing conditions is important to avoid ice buildup and to initiate electrochemical reactions. For this purpose, the heater position should provide good thermal connectivity with low heat loss and have a compact layout compatible with other system elements.

Most commonly, heaters are mounted externally around the stack casing or near end plates. This enables heat to transfer by conduction or convection and gradually warms the stack inside. Such a method is straightforward and modular, but fails badly in harsh, cold climates because heat transfer is less efficient and requires more energy input. Alternatively, the placement of heaters internally along gas flow paths into which hydrogen and oxidant gas are introduced leads to better heating results, and by pre-heating the gases, the system achieves temperature uniformity, along with membrane hydration, in response to the catalyst layers. This setup allows faster start-ups with lower energy demands, but gas flow and temperature must be controlled to avoid unwanted condensation/uneven heating. More advanced designs use flexible heaters or thin films directly inside the bipolar plates for heating of active electrochemical zones, which have low start-up lag and good integration with the stack control algorithms. In some designs, the two (internal and external heaters) are combined to achieve a compromise between efficiency, performance, and system complexity. In a successful configuration, distributed temperature sensors continuously monitor the stack to control heater operation, avoid overheating, and keep the PEMFC operating in its optimal thermal window of about 70–80°C. Ultimately, smart heater placement in combination with dynamic thermal management and responsive control strategies provides cold start capability with membrane protection and improves system durability (Longoria et al., 2021).

1.5.2 Thermal Coupling and Distribution Techniques

Efficient thermal coupling and distribution strategies are very important for hydrogen fuel cell systems, in particular configurations with catalytic heaters for cold start support, and these strategies are very critical for system efficiency, stability, and long-term operation life of the system when operating conditions are changing. Good thermal control increases the electrochemical reactions of fuel cells at freezing temperatures. This avoids risks of uneven heat distribution, which gradually damages the internal components of the system. Thermal coupling directs heat to another subsystem. For example, heat flow management between the active component and thermal buffers, and heat transfer from the catalytic heater to the stack. Meanwhile, thermal distribution disperses this heat to all required areas.

For these goals, advanced heat exchangers are used. These devices transfer heat from the coolant loop to the stack and from heated hydrogen to intake gases. With their high surface area/volume design and minimal thermal resistance, they are very adaptable to changing thermal demands. And passive systems like heat pipes and loop heat pipes can replace active pumping. They employ phase transitions and capillary forces to transfer heat rapidly around the stack and other modules in small-area fuel cell applications. PCMs and thermochemical storage media are thermal buffers. They store heat during regular operation and let heat out during start-up to maintain a constant internal temperature. On the air side, forced convection and pre-heated airflow distribute thermal loads fairly in open cathode designs. The biggest challenge is to reduce temperature differences across fuel cell stacks. Uneven heat distribution leads to catalyst degradation, stress on the membrane, or poor electrochemical performance. This leads to thermal integration at the system level. To achieve this, it should manage flow dynamics, intelligent control mechanisms, and material compatibility while controlling costs. As the demand for stable low low-temperature hydrogen fuel cells grows, finely tuned thermal coupling and distribution schemes are still required to fabricate scalable, reliable power solutions (Liu et al., 2024).

1.5.3 Pre-Heating of Air and Coolant Loops

Warming up air and coolant loops in advance is an important strategy for cold start management in hydrogen fuel cell systems under freezing conditions. This avoids performance risks of low temperatures. At very low temperatures, ice build-up, membrane dehydration, and sluggish electrochemical reactions seriously impair stack performance, and pre-heating compensates this by increasing the temperature of reactants and thermal fluids before start-up. This increases efficiency and system responsiveness while defending components from mechanical and chemical stress. In PEMFCs, which normally operate between 60°C and 80°C, a cold start without sufficient thermal preparation delays activation and shortens the fuel cell's life.

Some systems redirect exhaust gases or use dedicated heat exchangers to transfer warmth to the entering air stream before the cathode. This speeds up oxygen reduction but maintains proper humidity. It is important that this does not dry out the membrane. In some configurations, electric heaters initially provide thermal support during start-up and gradually fade out as the stack generates enough waste heat to maintain temperature. For the coolant loop, pre-heating always involves a small, closed thermal circuit equipped with resistive heaters or catalytic heat sources. In this setup, warmed coolants are circulated through the stack to achieve uniform heating of bipolar and end plates while reducing temperature differentials. More sophisticated designs employ thermal energy storage Materials like PCMs. This stores heat from earlier operations and releases it during the next start-up. This passive system saves energy consumption while preserving system readiness during intermittent use. Combining air and coolant pre-heating in a single thermal control strategy shortens start-up times, preserves the stack, and reduces wasted energy. Thus, good pre-heating techniques are needed for cold start reliability. This allows for the wide deployment of hydrogen fuel cells in different climates (Melnik et al., 2024).

1.5.4 Control Synchronization with Fuel Cell ECU

In hydrogen fuel cell systems – especially those designed for vehicles – the complete synchronization of catalytic heaters with the FCCU is critical to the performance, safety, and thermal efficiency during start-up and shutdown. This coordination helps in making the system more robust. Essentially, the central brain of the system, the FCCU, processes real-time data coming from many sensors while controlling actuators, pumps, valves, compressors, and even the catalytic heater itself. This synchronization is important in cold environments where the heater is turned on very slowly to warm the stack before electrochemical reactions can begin.

The FCCU continuously monitors ambient and stack temperatures, system pressure, hydrogen levels, and electrical demands for precise control. Using predefined cold start protocols, it switches on the catalytic heater only when temperature thresholds are reached, thus ensuring that energy is used only when pre-heating is necessary. At the desired thermal conditions, the FCCU controls hydrogen and oxidant flows, starts coolant circulation, and stabilizes the voltage output of the stack. This prevents thermal shock, limits membrane wear, and extends catalyst life, which are risks common when cold starts are poorly coordinated. The FCCU also communicates bidirectionally with supporting systems such as battery management units and powertrain electronic control units. This link allows intelligent energy sharing, so batteries or super capacitors can provide power for early heater activation/initial ramp-up of fuel cells. For fault handling, the FCCU performs continuous diagnostics, identifies abnormal heating patterns, monitors heater performance, and initiates safe shutdowns or fallback protocols if required. On advanced platforms, adaptive control strategies, predictive algorithms adjust heater timing, power timing, and power management according to the changing conditions. The catalytic heater synchronization with the FCCU also provides cold start reliability, extends system life, and conforms to safety standards in real-world operations (Dong & Jiang, 2024).

1.6 PERFORMANCE METRICS AND EFFICIENCY CONSIDERATIONS

Assessment of hydrogen fuel cells' efficiency and performance, generally under cold start conditions with catalytic heaters, requires a multi-layered short-and long-term operation stability evaluation. This approach allows a detailed understanding of system behavior. Catalytic heaters are useful to jump-start reactions and keep the membrane hydrated, but they represent an initial energy burden that must be accounted for in overall efficiency assessments and key performance indicators such as voltage stability and current density. Power output at different thermal loads tests the stack's resilience during cold start cycles. Transient characteristics such as warm-up duration and voltage delay are critical for mobile or emergency power systems, which require quick start-up.

External variables like humidity, ambient temperature, and the purity of hydrogen fuel affect system efficiency. Similarly, internal issues like reaction kinetics and electrode design are considered. In cold conditions, poor membrane hydration/ice formation decreases the effective electrochemical area. That causes higher activation losses and lower power output, while the heat supplied by catalytic heaters is carefully controlled to avoid hot spots damaging the membrane or uneven catalyst wear. In the larger energy consumption profile, the heaters' own efficiency is often taken into account as hydrogen conversion rates and heat output per unit energy input (Aminudin et al., 2023). Early in start-up, fuel utilization and mass transport are constrained by low gas diffusivity and possible water buildup in diffusion layers. For system response tuning, advanced simulation models are used to predict thermal–electrochemical interactions and control strategies. Cold start frequency and duration affect durability metrics like voltage decay rate and life cycle. Efficiency maximization goes beyond material and catalytic activity improvement. It demands efficient waste heat recovery, optimized start-up procedures, and close coordination among the thermal and electrochemical components of the system.

1.6.1 Heat Generation Rate and Uniformity

In hydrogen fuel cell systems, especially those based on PEMFCs, both rate and spatial spread of heat generation influence the system efficiency, structural reliability, and component lifetime, and this interplay affects the operational outcomes. Since the electrochemical reaction of hydrogen to produce electricity and water results in heat loss of about 45–60% of input energy, this thermal output can be attributed to more than one source – entropy from the redox process, resistive (Joule) heating due to thermal effects, internal electrical resistance, and catalytic losses due to phase transitions like gas diffusion layers or water vapor condensing inside the membrane.

In a fuel cell, the heat produced is unevenly dispersed, which makes thermal regulation difficult. The cathode side is generally warmer than the anode due to accelerated reaction activity and oxygen reduction reactions. This causes localized heat buildup and may disturb the membrane hydration. These irregular gas flow paths, the stack orientation (horizontal or vertical), and the difference in contact resistance across cells and layers contribute to this uneven heating. Such thermal disparities without proper management cause membrane drying, catalyst degradation by sintering, localized thermal strain, and even structural layer separation, affecting performance and shortening operation life. Against such risks, advanced thermal management strategies are applied to spread heat more uniformly and control the flux more generally. They include defining reactant and coolant flow fields precisely, channel geometry optimization for fluid balance, and incorporating high thermal conductivity materials into bipolar plates and stack end structures. Actively controlled cooling systems based on deionized water and glycol solutions are coupled to sensors and controllers to adjust flow rates dynamically in real time (Malik et al., 2023). For thermal uniformity purposes, novel coolants, heat pipes with pulses, and thermally enhanced composite membranes are gaining ground. Such an approach leads to optimal fuel cell performance at steady and fluctuating conditions with the right balance between the heat produced and the heat dispersed, especially in cold climates where system activation requires high start-up temperatures.

1.6.2 Hydrogen Consumption Efficiency

Hydrogen consumption efficiency is a dynamic concept in fuel cell systems that goes beyond electrochemical reactions within a stack. Although hydrogen fuel cells can change chemical energy to electricity at efficiencies typically between 40% and 60%, this numerical range does not represent the energy use in cold start situations, where extra hydrogen is needed for thermal support. The catalytic heater, for example, uses hydrogen independent of power production. Although they are important for cold climates, reliable ignition, these heaters temporarily divert H_2 from electricity generation, thus reducing utilization efficiency at start.

Several internal factors determine how efficiently the stack uses hydrogen: current density, temperature conditions throughout the cell, and membrane hydration level. At lower current loads, fuel cells tend to work more efficiently but with lower power output. But when power demands are high, efficiency decreases dramatically due to increased resistance-related (ohmic) losses and constraints in mass transport. Hydrogen purging also prevents hydrogen loss from the anode side if not controlled by recirculation or other recovery systems. The full hydrogen life cycle should be considered for understanding the broader energy performance from well to wheel efficiency: compression, production, distribution, and final use on board the vehicle. Hydrogen sourcing via electrolysis using renewable energy is cleaner but relatively inefficient, with energy conversion rates often below 60%. With transportation and storage losses accounted for, the net hydrogen energy available to the fuel cell drops dramatically. Yet, fuel cell vehicles outperform traditional internal combustion engines in overall efficiency and produce far fewer emissions (Alirahmi et al., 2021). Increasing hydrogen utilization requires system-level strategies such as thermal energy flow integration, stack insulation to minimize heat loss, hydrogen recirculating loops, and frequency and timing optimization of purge cycles. Batteries or supercapacitors are optional hybrid fuel cell configurations for smart energy management in variable load conditions. Such adaptations save unnecessary hydrogen consumption and improve system operation efficiency in various real-world driving scenarios.

1.6.3 Start-Up Time Reduction Benchmarks

Reducing the start-up time of hydrogen fuel cell systems is an important task in transportation and emergency backup systems where an immediate power supply is required. The most rapid activation time of the Fuel cell technologies, PEMFCs, is between 1 and 10 seconds when tuned. This high conductivity of polymer electrolyte membranes and low operating temperatures make them an excellent choice for portable and vehicular applications. Meanwhile, SOFCs that operate at temperatures above 700°C take from several minutes to hours to reach steady state due to thermal ramp-up and material stabilization. It is not a fixed metric starting time but rather a function of several design and operational parameters such as thermal mass of the system, specific heat capacities of its components, insulation quality, and heat production and distribution efficiency. As an important auxiliary system, catalytic heaters greatly reduce the PEMFC cold start durations by rapidly warming up the membrane and catalytic layers. Targeted or localized heating techniques are combined with finely tuned airflow management and preconditioning of the coolant system to minimize the thermal delays typical of low-temperature start.

New performance benchmarks for reducing start-up times are being developed in materials science, and nanostructured catalysts and low-resistance bipolar plates, for example, improve reaction kinetics and energy dissipation during warm-up phases. Intelligent control systems and predictive thermal models also allow smarter, anticipatory heating strategies to further shorten activation time. Hybrid configurations with batteries/supercapacitors allow initial load handling during fuel cell system ramp-up. Optimal starting time increases the practical usability and energy performance of hydrogen fuel cells. Also, it increases their commercial viability in high-speed, real-world applications where consistent, instant power delivery is required (Hami & Mahmoudimehr, 2023).

1.6.4 Trade-Off between Heater Size and Power Output

In hydrogen fuel cell systems, engineers have to balance heater size with power output, and this is particularly important in cold start scenarios requiring compact system architecture. The larger fuel cell stacks generally have a larger active surface area. The Dieser design allows operation at low current densities while maintaining uniform voltage across each cell. Moreover, the heat produced with increasing power generation is proportionately increased due to electrochemical process inefficiencies. Managing this additional heat requires more sophisticated or larger thermal systems, which increase the system weight, size, and cost.

Catalytic heaters used to pre-heat the stack before ignition require appropriate scaling with respect to the stack thermal mass and expected heat loss; oversized heaters result in higher hydrogen consumption and wasted energy; undersized heaters may delay system readiness, an issue that is acute in freezing conditions. Such a balancing act is especially critical in automotive or portable applications where weight reduction and component fitting in tight spaces remain a priority, while larger heaters require additional safety features to prevent localized overheating hot spots or hydrogen leakage. Therefore, to balance this delicate balance, engineers are increasingly relying on improved heat exchanger designs, modular heater configurations, and thermal energy storage solutions that can expand or contract in response to the changing power demand. Continually improved membrane technology, like modern low-humidity membranes, eliminates external humidification and dependence on large heating systems, allowing smaller auxiliary units. Furthermore, the use of hybrid energy systems, including batteries or ultracapacitors, spreads the initial start-up power load more evenly over the heater, and finally, the practical effectiveness of fuel cell systems is dependent on the management of heater size–power output relationship to achieve performance targets while preserving reliability, efficiency, and design compactness (Dezhdar et al., 2023).

1.7 SAFETY CONSIDERATIONS AND RISK MITIGATION

Catalytic heaters for hydrogen fuel cell systems require a holistic approach to safety because hydrogen is reactive and temperatures are high in catalytic reactions. These heaters operate on flameless hydrogen oxidation of coated surfaces of palladium or platinum, which produces heat without burning open flames and is suitable for cold start conditions. But despite these benefits, this method introduces some hazards like hydrogen leaks, excessive heat buildup, and material endurance challenges, and the interaction of hydrogen with various system components under fluctuating temperatures and pressures will probably introduce pressure instability, hydrogen embrittlement, and localized thermal stress complications, which, if not properly managed, may damage the system.

Attenuating these risks begins during design, and safe system architecture places heaters away from critical electronics and hydrogen storage areas to limit the impact of malfunctioning. Passive as well as active ventilation solutions are needed to prevent leaked hydrogen accumulation in enclosed or low-airflow environments, and thermal barriers, insulation layers, and real-time temperature monitoring are required to limit heat exposure and prevent accidental ignition or runaway thermal events. Modern automation is required, since control units that continuously monitor temperature, pressure, and hydrogen concentrations allow early issue detection/shutdown procedures when needed. A second fundamental aspect is material selection, and hydrogen-exposed components are made of metals and alloys that resist embrittlement and thermal cycling. Regular system inspections are necessary – especially for microleaks developing at seals, valves, and joints – and training of personnel operating and maintaining the systems is also necessary. Proper leak detection instruction, heater start-up protocols, and emergency responses ensure operational safety and conformity with international safety standards like ISO 19880 and IEC 62282, ensuring conformity with industry standards for hydrogen system design and usage. Through finely designed, monitored, and disciplined safety practices, catalytic heaters are safely integrated into fuel cell systems, supporting the larger goal of hydrogen energy deployment reliably and at scale (Kumar et al., 2024).

1.7.1 Hydrogen Handling and Containment

The special physicochemical properties of hydrogen – extremely light molecular weight, broad flammability range, and high diffusion rate – make its management and containment in fuel cell systems extremely delicate. These characteristics make hydrogen an excellent energy medium, but present serious safety challenges if not managed properly. This is relevant for cold start applications with catalytic heaters, where hydrogen leaks may cause localized ignition hazards or worse system performance. Hence, safe containment throughout the hydrogen chain – from storage and transport to delivery and final use – requires stringent safety protocols at every step. Pressurized hydrogen storage from 350 to 700 bar uses carbon fiber-reinforced polymer composites to prevent hydrogen seepage. These tanks generally have safety devices like non-return valves, thermal relief devices, and rupture discs, which will produce counter-pressure surges due to temperature change. For cryogenic storage, counter-pressure MLI and vacuum-insulated vessels are used to prevent thermal ingress and hydrogen boil-off, and often require active venting and pressure-balancing devices. Meanwhile, solid-state storage technologies like metal or chemical hydrides provide safer alternatives by chemically binding hydrogen into their structures, but these methods add complexity to heat management. This requires precise thermal regulation strategies for hydrogen absorption and release.

Distribution infrastructure should meet material compatibility standards, whereas piping and tubing should be austenitic stainless steel or specially treated polymers resisting hydrogen-induced degradation. Connectors and seals selected for pressure cycling resistance and hydrogen permeation resistance increase system durability and safety. Leak detection is another important element; electrochemical or infrared sensors are used to detect leaks in closed areas where buildup may be dangerous. In addition to the technological measures, system operators are trained and standards like ISO 19880-1/SAE J2601 are strictly observed. These combined approaches make hydrogen handling secure and reliable in fuel cell technology applications (Alfasfos et al., 2024).

1.7.2 Temperature Monitoring and Control

Effective temperature monitoring becomes mandatory for hydrogen fuel cell systems and is particularly important in systems using catalytic heaters for cold start situations and where engineers must optimize performance and always increase system life expectancy while maintaining operational safety. Temperature influences many core functions and internal stack operations, like electrochemical reaction rates or proton conduction mechanisms. Designers for PEMFCs must keep operating temperatures between 60°C and 80°C, and they must maintain consistent membrane hydration and avoid catalyst degradation by strict thermal regulation protocols. At temperatures outside that range, systems show reduced voltage stability and power output, while overall fuel efficiency decreases significantly under stress.

For these thermal control problems, future hydrogen fuel cells adopt advanced thermal management techniques and robust hardware systems. These systems use closed-loop coolant circulation methods and resistive heaters and conductive materials to adapt quickly. Rapid environmental changes trigger automatic responses, and heat-regulating components work in close cooperation with predictive thermal algorithms. Sensors are placed inside bipolar plates, end plates, and coolant channels for real-time thermal readings. This data is collected continuously by the FCCU, which implements PID-based control loops/predictive logic for heater regulation.

Integration of catalytic heaters at cold start becomes complex and requires precise handling to avoid thermal overshoot. Such heaters require proper start-up management, and systems need safeguards against stack material degradation or stress. Accurate thermal input controls heater modulation and timing of active fuel cell operation. For extra operational assurance, sectors like aerospace or automotive install backup sensors, and engineers enable fail-safe redundancies. Overall, robust thermal management secures fuel cell durability and ensures consistent energy output to enable wide adoption of hydrogen in critical industries (Cavaliere, 2023).

1.7.3 Catalyst Poisoning and Deactivation Risks

Catalyst poisoning compromises the long-term durability of hydrogen systems and reduces performance consistency across fuel cells using platinum-based electrocatalysts. These catalysts are still required for the hydrogen oxidation and oxygen reduction reactions (HOR) of the electrochemical conversion. This high surface reactivity keeps platinum vulnerable to many contaminant species. Even trace impurities affect performance and block active catalytic sites in PEMFCs. The difficulty is compounded in cost-sensitive applications where catalyst loading is minimal and economic constraints will persist.

Carbon monoxide is one of the most potent poisons and adheres to platinum catalyst surfaces. CO adsorbs hydrogen directly, and this interference will impair the HOR rate and the system's responsiveness. In systems using reformate hydrogen, this problem is exacerbated as CO appears even at parts-per-million concentrations. H_2S is also a major threat and irreversibly deactivates catalysts by forming persistent compounds. Chloride ions attack platinum surfaces and come from degraded membranes or impure water sources during operation. Others, such as ammonia and volatile organics, collect inside systems and become degraded over time.

To prevent such catalytic failures, future systems will include advanced purification devices, and gas quality will meet international standards like ISO 14687. For contaminant control, they use palladium membrane filters, pressure swing adsorb units, and sulfur scrubbers. Researchers also prepared novel catalyst formulations or explored resilient materials such as platinum–ruthenium alloys or doped metal oxide supports. The effects of poisoning were minimized by operational tuning, and increasing gas flow or raising temperature reduces the residence time of toxic species. Additionally, surface recovery by voltage cycling and pulsed oxidation is achieved while maintaining catalyst activity. This combination extends the catalyst lifespan and assures uniform system performance and reliability in variable conditions of operation (Rathore et al., 2021).

1.7.4 Compliance with Safety Standards and Protocols

Respecting safety regulations is mandatory for large-scale acceptance of hydrogen technologies – this is especially true when catalytic heaters are integrated for fuel cell cold start operations, and these components require strict oversight under regulatory norms. As low ignition energy, wide flammability range, and high diffusivity remain properties of hydrogen, systems need robust safety frameworks. These requirements lead institutions such as the IEC, ISO, SAE, and national authorities to establish international codes. Collectively, these standards provide guidelines for safe hydrogen system design and operation. Hydrogen fuel cells and their auxiliary components, such as catalytic heaters, must comply with hydrogen quality, performance, and crash safety benchmarks like ISO 14687/IEC 62282/SAE J2578. These codes define safety requirements for leak limits and fire protection, and thermal resilience requirements for systems. To stop abnormal conditions, engineering teams use safety measures like optical or electrochemical hydrogen sensors and automated shutdown mechanisms. In addition to this, ventilation channels are designed to prevent the accumulation of leaked gas, and electrical components are certified for hazardous zone deployment. Catalytic heaters undergo stricter checks and present no ignition hazards when operating under specified thermal limits.

Compliance efforts span the full system, life-cycle phases, and extend beyond engineering phases to commissioning activities. Installation protocol is based on regulatory blueprints; operators conduct periodic inspections and maintenance reviews for safety integrity. Final phases follow guidelines, too, and developers use risk evaluation methods like FMEA, HAZOP, and QRA. They identify system hazards, and developers implement countermeasures to avoid accidents proactively. Technical personnel are trained in operation protocols and emergency protocols. Normative supporting documentation is used, and system managers adopt SOP and MOC frameworks that drive procedural as well as structural changes. Internationally recognized compliance norms will protect human life and equipment integrity, and build public trust and industrial confidence simultaneously. This public assurance enables market adoption in transport systems and stationary energy platforms. All these efforts together lay the foundation for a safe, large-scale implementation of hydrogen systems worldwide (Kumar et al., 2024).

1.8 RECENT DEVELOPMENTS AND FUTURE PROSPECTS

A new hydrogen fuel cell technology raises operational standards, and cold start systems are improved by catalytic heater integration. Such technological advances lay the foundation for dependable and efficient hydrogen energy deployment and enable wider sectoral adoption. Innovation is increasingly focused on thermal regulation procedures optimization and cold start responsiveness optimization while maintaining economic viability and energy efficiency. Auxiliary heating elements are redesigned according to economic and power considerations, and system engineers integrate them in control modules. Catalytic heater configurations become streamlined and responsive, and stack-level controller integration improves hydrogen utilization rates.

System refinements reduce cold start times and improve operational reliability in low-temperature environments worldwide. Materials innovation keeps pace with research results, resulting in novel high-performance polymers and tough catalysts that can tolerate thermal transitions without losing performance consistency. Upgraded composite bipolar plates are standard, and new insulation technologies allow thermal control in catalytic houses. Hydrogen-permeable membranes help in safe combustion environments, and system safety protocols evolve accordingly for better operational resilience. Meanwhile, AI-based thermal modeling techniques optimize real-time heater diagnostics and predictive thermal balancing methods.

These innovations fine-tune energy usage profiles and improve the long-term dependability of start-up systems under stress. Researchers use machine learning to predict heat distribution profiles and calibrate heater parameters automatically, preserving thermal uniformity. Future hydrogen fuel cell cold start modules will adopt hybrid thermal strategies combining latent heat storage, heat recirculation systems, and phase change materials for better performance and emissions reduction. All these trends become stronger as environmental mandates get stricter and energy systems must reach zero-emission benchmarks in transport, grid backup, and industrial use cases.

In parallel with renewable cost declines, sustainable hydrogen sources are adopted in fuel cell systems. Meanwhile, this technological expansion is dependent on design bottlenecks being removed and critical system components being standardized. Reducing life-cycle costs becomes a defining goal, and global stakeholders prioritize infrastructure investment in response. As hydrogen becomes an important component in the decarbonization pathways, coordinated international frameworks are needed, and global cooperation will determine how hydrogen cold start integration will develop in the future (Moretto & Quong, 2022).

1.8.1 Advanced Catalyst Materials and Nanocoatings

Advancements in catalyst materials and nanocoatings represent a new paradigm for enhancing affordability, reliability, and cold start efficiency of hydrogen fuel cell systems. These systems benefit especially in low-temperature operation and rely on catalytic performance to initiate power delivery. Conventional platinum-based catalysts are still effective in electrochemical reactions and performance outputs, but are limited by high cost factors and operational degradation over repeated start–stop cycles over time. To address these material limitations, researchers investigate new nanostructured catalyst geometries like core-shell nanoparticles and hollow nanospheres.

Allocated nanoclusters with structural integrity are synthesized to optimize atomic configurations and catalytic reaction kinetics. Those morphologies increase both surface area exposure and electron mobility simultaneously and reduce the dependency on expensive platinum. Lower platinum loading leads to better performance outputs, and researchers maintain catalytic activity for long periods (Genovese et al., 2023). Platinum efficiency also improves, and overall cost burden can be lowered for broad deployment as well. Attention now turns to alternative catalyst formulations, avoiding precious metals and introducing low-cost materials like M-N-C (metal–nitrogen–carbon) complexes. These less expensive options react well in oxygen reduction reactions tuned by chemical structure. Metal–organic-framework-derived carbon backbones provide stability, and doping strategies with sulfur, nitrogen, or boron modify the electronic properties effectively. All these changes increase the number of catalytically active sites and confer long-term catalytic durability.

Carbon-based conductive supports like graphene layers and carbon nanotubes are included for electron transfer under dynamic conditions. These supports also increase structural strength, and the operational stress and thermal instability tolerance are reduced. Nanocoatings protect catalyst surfaces from corrosion, oxidation, and physical degradation. Amorphous carbon films, oxide coatings, and self-assembled monolayers resist freeze–thaw cycling and voltage-induced stress. They maintain catalyst integrity and improve long-term reliability while regulating water retention and gas flow behavior. Customized pore distributions and adjustable hydrophobicity allow mass transport efficiency in cold conditions. Fabrication innovations will be electrospinning, sol-gel processes, and atomic layer deposition, allowing precise morphological control. Such control enables nanoscale shape and distribution of catalysts and improves electrochemical output and operational stability. Such technical progress supports economic targets of reduction of production costs and extension of fuel cell component life cycles. All such developments support the long-term goal of building scalable and robust hydrogen systems. Future fuel cell architectures combine cold start resilience with high performance expectations, and engineers meet climate adaptive demands via thermal innovation. These evolving solutions enable their deployment in transport, stationary, and industrial sectors, and their expansion drives clean hydrogen adoption.

1.8.2 Hybrid Heating Systems with Battery Assistance

In hybrid heating solutions, catalytic hydrogen fuel cells are combined with a battery backup system, and this is an important development in cold start technologies. Such combined systems have higher thermal responsiveness and energy efficiency and work reliably in various environmental conditions. Designers combine consistent heat output from catalytic heaters via hydrogen oxidation on catalyst surfaces with fast-acting battery elements. Such systems provide balanced heating and are ready for cold starts when required, while maintaining operational equilibrium.

Battery modules start rapidly at system start-ups, with components such as sensors, thermal controllers, and pre-heating systems at system initiation. At the same time, catalytic heaters start warming the fuel cell stack and activating electrochemical reactions. This dual mechanism helps temperature-sensitive elements like the MEA reach optimal conditions very quickly. This will result in a minimum start-up delay, low thermal shocks, and critical parts will work safely during the cold start cycle. In this hybrid configuration, catalytic heaters can work with lower hydrogen input, and energy demands can be shared effectively. In such designs, fuel efficiency improves, and hydrogen consumption is reduced while power output remains constant. This allows the use of compact and cheap fuel cell stacks without performance compromise during start-up. In addition, surplus energy is stored in a battery, either from regenerative brakes or solar arrays in static setups, and is diverted later for thermal or electrical assistance during peak demand periods.

Advanced power management systems control modules for coordinated operation and give priority to the battery during high-demand events. Once thermal conditions become stable, the control software transfers the energy load to the fuel cell efficiently. This even exchange of energy increases thermal utilization efficiency without severe operation stress on components. The life span of both system elements will increase, and reliability will remain high during cold-weather deployments in remote areas. For isolated or off-grid applications, hybrid systems provide consistent and rugged heating in emergency shelters, military bases, or communications facilities. Such thermal management solutions are resilient, reliable, and provide good energy distribution. Adoption rates increase as battery and hydrogen infrastructure become more accessible and economically feasible worldwide. Hybrid heating systems are a scalable, sustainable, and technically reliable solution to energy needs in cold environments (Waseem et al., 2023).

1.8.3 Automotive and Aerospace Applications

Hydrogen fuel cell technology has an increasing momentum both in the automobile and aerospace industries, and it is designed for higher power density, zero emission capability, and long-duration, high-power operating conditions. In the automobile industry, FCEVs compete with battery-powered vehicles and are increasingly adopted in long-distance and commercial transit sectors. FCEVs enable short refueling times of less than five minutes and long ranges of over 500 kilometers. These are attractive for use in public buses, delivery fleets, and passenger vehicles under fluctuating, cold environmental conditions. Cold start performance is improved by catalytic heaters incorporated to overcome a major limitation of battery-based systems.

Automobile makers react to this industry momentum and invest more money in fuel cell technologies and innovation pipelines. Major players like Toyota, Hyundai, and Honda introduce commercial FCEV models like the Mirai, Nexo, and Clarity, respectively. These vehicles have advanced hydrogen storage at 700 bar and are hybrid-integrated with regenerative brakes. Thermal management systems also operate at stack-level precision and improve energy efficiency in different climates and operation profiles. Widespread adoption remains challenging, and the lack of hydrogen refueling infrastructure prevents growth, together with the high cost of platinum-group metal catalysts and limitations in green hydrogen production economics.

In the aerospace industry, hydrogen fuel cells are increasingly considered as alternatives to conventional auxiliary power units (APUs). These have replaced traditional battery banks in manned and unmanned aircraft and provide steady-state electrical output onboard. They are compact in size and high in energy reliability, suitable for avionics, climate control, and emergency backup systems. Pioneering efforts such as Airbus ZEROe/Boeing hybrid-electric aircraft advance research and validate fuel cell applications for propulsion/auxiliary support tasks. UAVs will really benefit from such systems, and project examples like Helios and Phantom Eye show better endurance in real conditions than conventional batteries.

Despite such benefits, aviation deployment presents challenges that must be resolved before certification. Hydrogen must be cooled cryogenically, and system components must be able to withstand high vibration in flight. Meeting aviation standards entails stringent fire safety compliance/airworthiness validation processes within jurisdictions and aircraft classes. Material science technology advances and onboard fuel reforming methods improve with compact system architecture. With catalytic heating for low-temperature start-up performance, hydrogen systems are extended. All these innovations position fuel cells at the heart of future sustainable mobility solutions for ground and air vehicles.

1.8.4 Power Solutions and Emerging Trends

Hydrogen fuel cells are transformative technologies in the global energy framework, and their catalytic heating system provides cold start capability in all environments. Such systems produce electricity from electrochemical reactions of hydrogen with oxygen but release only water vapor as a product. Their environmental compatibility positions them at the cutting edge of zero-emission technologies and sets new efficiency standards. Such cells outstrip traditional combustion engines in applications requiring instant power and environmental resilience and are applicable to transport, marine, and off-grid power applications.

Power generation systems are developing rapidly in all sectors, and hydrogen fuel cells transform operations in commercial and heavy transport. For heavy-duty transport segments like intercity buses, commercial trucks, and rail systems, fuel cells will provide extended range and fast refueling and will address the drawbacks of battery-based options. For distributed energy applications in telecom towers/remote medical sites/industrial installations, fuel cells will offer modular configurations with advanced thermal control capabilities. The maritime sector, too, is adopting fuel cells as it responds to emission regulations and replaces marine diesel engines with cleaner technologies. Renewable-powered electrolysis techniques change green hydrogen production and supply chains. The transition allows renewable electricity integration into chemical storage formats while reducing the life-cycle carbon footprint. Parallel developments in catalyst engineering, fuel cell architecture, and solid-state storage methods increase operational efficiency and lower capital costs. Hybrid configurations combine batteries with fuel cells for flexibility under dynamic loads. These work better in cold environments and can react to changing energy demands.

Fuel cells are affected by digital transformation, and artificial intelligence tools allow real-time diagnosis. IoT technologies provide predictive maintenance functions, and data analytics give system-level visibility. Infrastructure expanding in Europe and Asia means adoption is spreading globally. Still, some barriers exist in terms of cost inefficiencies and scalability, and infrastructure development is required for mass adoption. Yet, continued technological innovation and coordinated policy support define the future of hydrogen adoption across all industrial domains.

1.9 CONCLUSIONS

Integrating catalytic heaters in hydrogen fuel cell systems is a reliable and low-power strategy to overcome the cold start challenge and helps to improve start-up performance. They produce essential thermal energy by controlled oxidation of hydrogen, and they speed up the process of reaching the optimal operating temperature in the fuel cell stack, reducing delays in cold conditions. This thermal support significantly improves fuel cell responsiveness and system efficiency and reliability, and is useful in freezing or very variable climates. This chapter describes the trade-off between catalytic heater size and heat output capacities, insulation quality, durable construction materials, gas flow regulation mechanisms, embedded safety systems, and integrated thermal management strategies in various applications and operating scenarios. Particular attention was paid to catalyst stability over long durations and precise hydrogen management, and these requirements are representative of the stringent engineering constraints and regulatory frameworks that these systems must meet consistently. Among technological advances are the creation of battery-assisted hybrid heating solutions, new generation catalyst materials characterized by high thermal resilience, and diagnostic technologies for real-time monitoring of performance indicators. These advances push cold start performance forward and contribute to system durability and economic viability, which is necessary for adoption in vehicle/aircraft and stationary fuel cell sectors. These are flexible enough to satisfy diverse energy needs and reflect the current global focus on clean energy technologies and hydrogen infrastructure expansion efforts. They become the backbone of larger trends such as the introduction of green hydrogen production, smart thermal regulation approaches, and safe and sustainable infrastructure construction. Ultimately, catalytic heaters are an enabling technology to ensure cold start dependability and the industrial application scope of hydrogen fuel cells, and their continuous improvement is a prerequisite to realizing hydrogen's full potential as a clean and flexible energy source in diverse environmental and application contexts. Their evolution reinforces long-term energy transitions toward sustainability, and they support global climate objectives with scalable and competitive power alternatives that reduce emissions but maintain performance.

REFERENCES

Alfasfos, R., Sillman, J., & Soukka, R. (2024). Lessons learned and recommendations from analysis of hydrogen incidents and accidents to support risk assessment for the hydrogen economy. *International Journal of Hydrogen Energy*, 60, 1203–1214.

Ali, M. M., Hussain, A., Song, R. H., Khan, M. Z., Park, S. J., Ishfaq, H. A., … & Lim, T. H. (2024). Beyond traditional fuel cells: Development and a comprehensive analysis of mechanically robust metal mesh-supported solid oxide fuel cell. *Ceramics International*, 50(21), 41028–41038.

Alirahmi, S. M., Assareh, E., Chitsaz, A., Holagh, S. G., & Jalilinasrabady, S. (2021). Electrolyzer-fuel cell combination for grid peak load management in a geothermal power plant: Power to hydrogen and hydrogen to power conversion. *International journal of hydrogen energy*, 46(50), 25650–25665.

Aminudin, M. A., Kamarudin, S. K., Lim, B. H., Majilan, E. H., Masdar, M. S., & Shaari, N. (2023). An overview: Current progress on hydrogen fuel cell vehicles. *International Journal of Hydrogen Energy*, 48(11), 4371–4388.

Cavaliere, P. (2023). Safety issues and regulations. *Water electrolysis for hydrogen production* (pp. 729–791). Springer International Publishing.

Dezhdar, A., Assareh, E., Agarwal, N., Keykhah, S., Aghajari, M., & Lee, M. (2023). Transient optimization of a new solar-wind multi-generation system for hydrogen production, desalination, clean electricity, heating, cooling, and energy storage using TRNSYS. *Renewable Energy*, 208, 512–537.

Dong, L., & Jiang, P. (2024). Improved super-twisting sliding mode control strategy in permanent magnet synchronous motors for hydrogen fuel cell centrifugal compressor. *Heliyon*, 10(2), e24181.

Ertekin, D., Baltaci, K., Toprak, M. Z., Çelebi, M., Özden, M., & Siano, P. (2024). *The design and practical realization of an adaptable grid integrating hydrogen fuel cell setup with a fuzzy-logical controller-based SVPWM boosted inverter. IEEE Access*, 12, 123489–123513.

Ferrara, A., & Hametner, C. (2021). Impact of energy management strategies on hydrogen consumption and start-up/shut-down cycles in fuel cell-ultracapacitor-battery vehicles. *IEEE Transactions on Vehicular Technology*, 71(6), 5692–5703.

Genovese, M., Cigolotti, V., Jannelli, E., & Fragiacomo, P. (2023). Current standards and configurations for the permitting and operation of hydrogen refueling stations. *International Journal of Hydrogen Energy*, 48(51), 19357–19371.

Guan, W., Chen, J., Chen, L., Cao, J., & Fan, H. (2023). Safe design of a hydrogen-powered ship: CFD simulation on hydrogen leakage in the fuel cell room. *Journal of Marine Science and Engineering*, 11(3), 651.

Hami, M., & Mahmoudimehr, J. (2023). When to switch from heat-up to start-up in the warming-up process of a solid oxide fuel cell: A numerical study and multi-objective planning. *Journal of Power Sources*, 585, 233656.

Hossain, M. A., Islam, M. R., Hossain, M. A., & Hossain, M. J. (2023). Control strategy review for hydrogen-renewable energy power system. *Journal of Energy Storage*, 72, 108170.

Kumar, S., Nanan-Surujbally, A., Sharma, D. P., & Pathak, D. (2024). Hydrogen safety/standards (national and international document standards on hydrogen energy and fuel cell). *Towards hydrogen infrastructure* (pp. 315–346). Elsevier.

Liu, S., Li, C., Zachman, M. J., Zeng, Y., Yu, H., Li, B., ... & Wu, G. (2022). Atomically dispersed iron sites with a nitrogen–carbon coating as highly active and durable oxygen reduction catalysts for fuel cells. *Nature Energy*, 7(7), 652–663.

Liu, J., Yang, F., Wu, Z., & Zhang, Z. (2024). A review of thermal coupling system of fuel cell-metal hydride tank: Classification, control strategies, and prospect in distributed energy system. *International Journal of Hydrogen Energy*, 51, 274–289.

Longoria, G., Lynch, M., & Curtis, J. (2021). Green hydrogen for heating and its impact on the power system. *International Journal of Hydrogen Energy*, 46(53), 26725–26740.

Luo, M., Zhang, J., Zhang, C., Chin, C. S., Ran, H., Fan, M., ... & Shuai, Q. (2022). Cold start investigation of fuel cell vehicles with coolant preheating strategy. *Applied Thermal Engineering*, 201, 117816.

Maiket, Y., Yeetsorn, R., & Kaewmanee, W. (2022). Hydrogen flow controller applied to driving behavior observation of hydrogen fuel cell performance test. *ACS Omega*, 7(43), 38277–38288.

Malik, F. R., Yuan, H. B., Moran, J. C., & Tippayawong, N. (2023). Overview of hydrogen production technologies for fuel cell utilization. *Engineering Science and Technology, an International Journal*, 43, 101452.

McKinlay, C. J., Turnock, S. R., & Hudson, D. A. (2021). Fuel cells for shipping: To meet on-board auxiliary demand and reduce emissions. *Energy Reports*, 7, 63–70.

Meda, U. S., Rajyaguru, Y. V., & Pandey, A. (2023). Generation of green hydrogen using self-sustained regenerative fuel cells: Opportunities and challenges. *International Journal of Hydrogen Energy*, 48(73), 28289–28314.

Melnik, D., Bürger, I., Mitzel, J., Käß, J., Sarkezi-Selsky, P., Jahnke, T., & Knöri, T. (2024). Energy efficient cold start of a polymer electrolyte membrane fuel cell coupled to a thermochemical metal hydride preheater. *Applied Energy*, 359, 122585.

Moretto, P., & Quong, S. (2022). Legal requirements, technical regulations, codes, and standards for hydrogen safety. *Hydrogen safety for energy applications* (pp. 345–396). Butterworth-Heinemann.

Nguyen, H. Q., & Shabani, B. (2021). Review of metal hydride hydrogen storage thermal management for use in the fuel cell systems. *International Journal of Hydrogen Energy*, 46(62), 31699–31726.

Pramuanjaroenkij, A., & Kakaç, S. (2023). The fuel cell electric vehicles: The highlight review. *International Journal of Hydrogen Energy*, 48(25), 9401–9425.

Qu, Y., Li, D., & Wang, R. (2024). Study on integrated thermal management system of hydrogen fuel cell vehicles based on heat pump air conditioning. *Journal of Cleaner Production*, 434, 139951.

Rathore, S. S., Biswas, S., Fini, D., Kulkarni, A. P., & Giddey, S. (2021). Direct ammonia solid-oxide fuel cells: A review of progress and prospects. *International Journal of Hydrogen Energy*, 46(71), 35365–35384.

Rossetti, I., & Tripodi, A. (2024). Catalytic production of renewable hydrogen for use in fuel cells: A review study. *Topics in Catalysis*, 67(19), 1286–1305.

Sultan, H. M., Menesy, A. S., Hassan, M. S., Jurado, F., & Kamel, S. (2023). Standard and quasi oppositional bonobo optimizers for parameter extraction of PEM fuel cell stacks. *Fuel*, 340, 127586.

Tang, Q., Li, B., Yang, D., Ming, P., Zhang, C., & Wang, Y. (2021). Review of hydrogen cross-over through the polymer electrolyte membrane. *International Journal of Hydrogen Energy*, 46(42), 22040–22061.

Tao, X., Zeng, Z., Gao, W., Yan, C., Liu, H., Sun, K., … & Wang, T. (2025). Experimental study of cold start of PEM fuel cell with non-uniform metal foam flow field. *Applied Energy*, 389, 125721.

Wang, Y., Pang, Y., Xu, H., Martinez, A., & Chen, K. S. (2022). PEM Fuel cell and electrolysis cell technologies and hydrogen infrastructure development–a review. *Energy & Environmental Science*, 15(6), 2288–2328.

Waseem, M., Amir, M., Lakshmi, G. S., Harivardhagini, S., & Ahmad, M. (2023). Fuel cell-based hybrid electric vehicles: An integrated review of current status, key challenges, recommended policies, and future prospects. *Green Energy and Intelligent Transportation*, 2(6), 100121.

Yu, L. J., Yang, X. H., Zhang, Z. X., Lin, Z. H., Jin, Z. J., & Qian, J. Y. (2023). Parametric analysis on throttling elements of conical throttling valve for hydrogen decompression in hydrogen fuel cell vehicles. *Journal of Energy Storage*, 65, 107342.

Zhang, J., Ricote, S., Hendriksen, P. V., & Chen, Y. (2022). Advanced materials for thin-film solid oxide fuel cells: Recent progress and challenges in boosting the device performance at low temperatures. *Advanced Functional Materials*, 32(22), 2111205.

Zion, N., Douglin, J. C., Cullen, D. A., Zelenay, P., Dekel, D. R., & Elbaz, L. (2021). Porphyrin aerogel catalysts for oxygen reduction reaction in anion-exchange membrane fuel cells. *Advanced Functional Materials*, 31(24), 2100963.

2 Electrolyte-Heating Circuits

2.1 INTRODUCTION

The electrolyte-heating in hydrogen fuel cells is very important during cold start conditions when ambient temperatures are below the fuel cell system operation temperature limit. The electrolyte is usually a proton-exchange membrane (PEM) that carries protons from the anode to the cathode but is impervious to electrons and gases. It must therefore maintain a temperature range within which it can conduct ions well. Below that range, the membrane is unable to conduct protons effectively, leading to poor fuel cell performance and even start-up failure.

Cold weather is problematic as water within the membrane or in adjacent components may freeze, thereby blocking ionic pathways and increasing internal resistance. This requires external or integrated heating means for warming the electrolyte prior to or during start-up. Electrolyte-heating gets the membrane to a temperature where proton mobility and electrochemical reactions can occur safely. It is addressed by heating techniques including resistive heaters embedded in or near the membrane assembly, induction heating, and advanced material-based systems like carbon nanotube (CNT) heaters. These methods deliver targeted heat into the electrolyte and thus achieve a uniform temperature rise. Good heating restores the conductive ion of the membrane and helps with water management, as ice melts and condensates do not build up. Fundamentals of electrolyte-heating also involve understanding the thermal properties of the membrane, like specific heat capacity, thermal conductivity, and temperature sensitivity of proton exchange capacity. Heating strategies should avoid hot spots and material degradation/energy wastage. So, properly managed electrolyte heat is required for cold environment operation of hydrogen fuel cells (Guo et al., 2024).

2.1.1 Role of Electrolytes in Fuel Cell Operation

Electrolytes are the main parts of hydrogen fuel cells – they enable the transportation of ions, mainly protons, from the anode to the cathode, yet stop the immediate passage of electrons through the cell. The electrolyte is in most transport and portable power fuel cells, PEM fuel cells, which are solid polymer membrane devices. This membrane carries the hydrogen ions produced at the anode when hydrogen gas is divided into electrons and protons. The electrons pass through an external circuit to supply electrical power, and the protons move through the electrolyte to the cathode to join with oxygen and return electrons to make water. Its efficiency in this ion transport process directly affects the performance of the fuel cell. The electrolyte should allow high proton conductivity but be impervious to gases like hydrogen and oxygen. This selective conductivity makes sure the electrochemical reactions happen

DOI: 10.1201/9781003657866-2

only at the chosen electrodes, and thus, the system stability and energy efficiency are preserved. Also, the electrolyte has to be chemically stable, mechanically durable, and thermally resistant under diverse operating conditions, especially in systems at fluctuating or extreme temperatures.

Water management is another critical role of the electrolyte. The membrane should always remain sufficiently hydrated for proper proton conductivity. When this is dry, ionic transport is inefficient; if it becomes too hydrated or subjected to freezing conditions, it will swell, crack, or become unsupported. Accordingly, the correct thermal and moisture balance is mandatory for safe operation. In summation, an electrolyte of a fuel cell is an active medium that controls ion transport, electrochemical reactions, and overall system reliability and durability. This helps convert the chemical energy to electrical power. Electrolytes are the main parts of hydrogen fuel cells – they permit ions to go between the anode and cathode and maintain the electrochemical responses necessary to make energy. Their properties affect the fuel cell efficiency, performance, and reliability. Table 2.1 summarizes critical functions of electrolytes and their role in fuel cell behavior (Shi et al., 2020).

Table 2.1 shows that electrolytes are more than passive components – they are active components contributing to the fuel cell performance and resilience under various environmental conditions.

2.1.2 Impact of Low Temperatures on Ionic Conductivity

Low temperatures negatively affect the ionic conducting capacity of the electrolyte in hydrogen fuel cells based on PEMs. The electrolyte carries protons from the anode to the cathode in fuel cell operation. This depends on temperature and water content in the membrane. The membrane remains hydrated at ideal operating

TABLE 2.1
Role of Electrolytes in Fuel Cell Operation

Function	Description	Impact on Fuel Cell
Ionic Conduction	Transports protons (H^+) or hydroxide ions (OH^-) across the cell	Ensures continuous current generation and electrochemical efficiency
Electrical Insulation	Prevents electron flow through the electrolyte	Maintains the correct flow of electrons through the external circuit
Chemical Stability	Withstands fuel and oxidant environments without degradation	Enhances fuel cell durability and performance
Thermal Management	Operates effectively within required temperature ranges	Supports stable operation and cold start capability
Mechanical Support	Acts as a physical separator between the anode and the cathode	Prevents short circuits and maintains structural integrity
Water Management (in PEMFCs)	Aids in hydration for proton conduction while avoiding flooding	Balances performance and longevity
Compatibility with Electrodes	Interfaces well with catalyst layers and gas diffusion layers	Ensures minimal resistance and optimal reaction rates

temperatures of 60–80°C. However, below-freezing temperatures introduce several problems that impair ionic conductivity severely. The key is reduced membrane hydration. Low-temperature water in the membrane may freeze and block proton conduction pathways. Because proton transport in PEM fuel cells occurs via water, proton mobility decreases dramatically without liquid water. This limits the rate of electrochemical reactions and reduces the output of the fuel cell. Frozen water also increases the internal resistance of the cell, making start-up difficult or impossible in cold weather.

A further problem is that the low-temperature stiffening of the polymer structure of the membrane may prevent the free motion of functional groups for proton exchange. This mechanical rigidity further decreases ionic conductivity. In addition, repeated freeze cycles degrade the membrane structure with repeated use, causing performance losses and durability problems. To overcome these issues, external heating systems are often integrated into the fuel cell stack to pre-heat the electrolyte before start-up. The electrolyte should be maintained within the temperature range that allows sufficient hydration and polymer flexibility to operate safely. Hence, the effect of low temperatures must be controlled to maintain fuel cell readiness for cold environments (Lu et al., 2022).

2.1.3 Thermal Activation Requirements

Thermal activation usually means warming the cell pieces – especially the electrolyte – to a temperature at which electrochemical reactions and ion transport are able to happen efficiently in hydrogen fuel cells. It becomes especially critical during cold start conditions when ambient temperatures are below the fuel cell operating range. The electrolyte is usually a PEM and must be at a certain temperature to allow proper proton conductivity while preserving hydration. The proton mobility within the membrane decreases sharply at lower temperatures, which prevents the cell from producing electricity efficiently.

Most PEM fuel cells have thermal activation thresholds between 50°C and 80°C. Below this range, the membrane polymer structure becomes less flexible, and the water inside may freeze at sub-zero temperatures. This freezing reduces the conductivity and may physically damage the membrane over a long period of time as well. Thus, thermal activation before/during start-up is mandatory to avoid performance degradation and structural damage. Effective thermal activation also ensures that electrochemical reactions at the electrodes occur safely. Low-temperature reaction kinetics are slower, and the internal resistance of the cell is greater. This leads to lower voltage output and more energy consumption to maintain even minimal power generation. By increasing the temperature of the electrolyte and surrounding components, thermal activation restores proton exchange, catalyst activity, and water management conditions. Heaters designed for thermal activation are usually resistive heaters, embedded films, or inductive elements, which quickly deliver heat to the electrolyte layer. These must be energy-efficient, safe, and fast-acting to meet thermal demands without excessive power loss. Thus, properly managed thermal activation is required for cold start reliability, energy waste minimization, and extension of service life of hydrogen fuel cells under various environmental conditions (Nguyen & Shabani, 2021).

2.1.4 Overview of Electrolyte-Heating Methods

In hydrogen fuel cells, electrolyte-heating methods consider cold start challenges by raising the temperature of the PEM to achieve good ionic conductivity, respectively. Low-temperature fuel cells fail to conduct protons well because the electrolyte has less membrane hydration and more frozen water. Overcoming this requires specialized heating methods to get the membrane to the optimal operating temperature range of 60–80°C before the cell starts generating power. Probably the most common is resistive heating with embedded electrical heating elements near or within the membrane electrode assembly (MEA). These elements change electrical energy to heat by resistance and directly warm the electrolyte! Resistive heaters are simple, inexpensive, and precise, and are therefore widely used for initial start-up heating. Another method involves the incorporation in the fuel cell layers of conductive and carbon-based heating films, such as CNT and graphene coatings. They are quick and uniform heating materials with low energy consumption and are suitable for compact and lightweight systems.

Inductive heating is also gaining ground in advanced fuel cell designs. This method produces heat by causing electrical currents in conductive materials to vibrate in electromagnetic fields. This gives fast responses and eliminates the need for direct contact heating elements, thus reducing material stress and increasing durability. Some systems circulate warm liquid or air around the cell stack for indirect heating. This is a slower method, but it is useful in automotive applications if waste heat from other systems is reused. Each method has advantages and disadvantages, and the choice is based on system size, application, required start-up time, energy availability, and cost. Finally, good electrolyte-heating is essential for cold climate fuel cell performance (Singh et al., 2023).

2.2 TYPES OF HEATING CIRCUITS

Many different types of heating circuits are employed in hydrogen fuel cells for cold start challenges, such as getting the electrolyte and other parts to optimum operating temperatures very fast. The choice of heating circuit depends on system size, energy demands, start-up time, and cost considerations. The most common are resistive heating circuits based on the Joule heating effect. This involves electrical current passing through resistive materials, generating heat transferred directly to the electrolyte or other surrounding structures. They are simple circuits, easy to integrate, and have precise temperature control, suitable for small fuel cell systems. Inductive heating circuits are more advanced. These systems apply alternating magnetic fields to induce electrical currents in conductive components, which then heat up due to internal resistance. The rapid and contactless heating of inductive heating reduces the wear of physical components. However, generally, more complex control systems and detailed design are needed for uniform heating without overheating sensitive areas.

A different approach involves embedded carbon-based heating elements like CNT films or graphene-coated layers. These are excellent thermal conductors, flexible, and lightweight materials. They can be incorporated directly in the MEA for uniform, energy-efficient, and fast heating. Some systems also employ fluid-based or air-based heating circuits in which pre-heated air/coolant is circulated over the fuel cell stack to warm it up. They are slower in response than direct electric heating, but effective in large systems and vehicles where waste heat from other parts is employed. Each heating circuit type has its own merits and drawbacks, and sometimes a hybrid approach is adopted to balance energy efficiency, start-up speed, system complexity, and long-term durability for different hydrogen fuel cell applications. The cold start problem of hydrogen fuel cells is solved by heating the electrolyte and cell stack to operational temperatures using various heating circuits. The heating circuit selection depends on power demand, heating speed/cost, and compatibility with fuel cell materials (Zhou et al., 2023). Table 2.2 lists the main heating circuit types, working principles, and key characteristics.

2.2.1 Resistive Heating Elements

Cold start problems in hydrogen fuel cells are among the most common solutions using resistive heating elements. Those elements work on the Joule heating principle: current changes electrical energy to thermal energy through a resistive material. The heat generated warms the electrolyte to its optimal operating temperature, in this case, the PEM. This is critical in cold conditions, where low temperatures decrease ionic conductivity dramatically and sometimes freeze water in the membrane, preventing the fuel cell from starting. Many resistive heating elements contain metal alloys, conductors, and carbon composites. These materials are selected because they can produce consistent heat reliably while withstanding chemical and thermal stress inside a fuel cell. Heaters can be embedded in bipolar plates, integrated into MEAs, or placed adjacent to the electrolyte layer. This near location distributes heat to the components that need it most, thus reducing energy loss and enhancing start-up performance. Resistance heating has the advantage of being simple to control. Temperature control is possible through adjustment of current flow to achieve uniform heating without localized overheating or material degradation. Also, resistive

TABLE 2.2
Types of Heating Circuits Used in Fuel Cells

Heating Circuit Type	Working Principle	Key Characteristics
Resistive Heating Elements	Converts electrical energy into heat via Joule heating in metal conductors	Simple design, easy control, widely used, moderate efficiency
Inductive Heating	Generates heat through electromagnetic induction in conductive materials	Rapid heating, non-contact method, requires precise control electronics
Microwave Heating	Uses microwave radiation to excite and heat polar molecules in the material	High-speed heating, selective targeting, may require shielding
Embedded Carbon Nanotube Heating Films	Conducts electricity through nanomaterials embedded in flexible substrates	Lightweight, efficient, uniformly distributes heat, and scalable fabrication

heaters can be manufactured in different shapes and sizes to suit various fuel cell designs and applications.

But resistive heating also requires proper power management to avoid excessive energy consumption during long, cold conditions. In modern systems, it is often combined with a smart controller and thermal sensors for energy saving. In general, resistive heating elements are still a practical and reliable way to achieve cold start success and to maintain performance in hydrogen fuel cell systems.

2.2.2 Inductive and Microwave Heating

Advanced thermal management techniques, such as inductive and microwave heating, are applied in hydrogen fuel cells to solve cold start problems by rapidly increasing the internal temperature, such as the electrolyte. Inductive heating and microwave heating produce heat by electromagnetic principles, in contrast to conventional resistive heating based on direct electrical current through resistive material.

Inductive heating produces an alternating magnetic field with a coil that produces eddy currents within nearby conductive materials like metal bipolar plates. These induced currents create internal resistance in the material, which heats up from within. This method is simple to use, contactless, and offers localized heating that can be controlled. There are no embedded wires, heaters in contact with the membrane and electrode layers, causing less physical degradation or electrical interference. Inductive systems are particularly applicable to compact fuel cell stacks where space is tight and thermal control is precise. By contrast, microwave heating uses high-frequency electromagnetic waves to heat dielectric materials directly. In response to microwave energy, polar molecules within the electrolyte or adjacent materials swish rapidly, generating heat by molecular friction. Such a method can heat non-metallic and composite materials uniformly and quickly, which are suitable for some advanced fuel cells. However, careful tuning avoids hot spots and damage because of uneven absorption.

Each method has a quick start-up capability as well as an improved energy efficiency over traditional heating methods. They are, however, also very complicated in design and require shielding and precise control systems for safety and uniformity. Despite these challenges, inductive heating and microwave heating are still attractive technologies for next-generation hydrogen fuel cells, especially for applications requiring fast and stable cold start performance in extreme environmental conditions (Sharifvaghefi & Zheng, 2022).

2.2.3 Embedded Carbon Nanotube Heating Films

CNT heating films embedded in hydrogen fuel cells are an advanced thermal management solution, especially for cold start problems. These films exploit the excellent electrical and thermal conductivity of CNTs for instantaneous, uniform, and low-energy heating directly in the fuel cell structure. Integrating the CNT films into the MEA or coating the bipolar plates with a surface coating will deliver localized heating where needed, enabling the PEM to reach the optimal operating temperature very quickly. CNTs are round nanostructures of carbon atoms in a hexagonal lattice, which are great conductors of power and good heat dispersers. Electrical current passing through the CNT film produces heat via resistive (Joule) heating. Since the nanoscale structure and large surface area mean this heat is distributed evenly, it avoids hot spots, which could damage sensitive fuel cell materials. CNT films are lightweight and flexible, thus allowing their integration into compact fuel cell designs without bulk or mechanical stress.

CNT heating films have a fast thermal response, which is one of their advantages. These films reach the desired temperature in seconds, reducing start-up time and system readiness in cold environments. The lower energy consumption of CNT heaters also contributes to energy efficiency, which is an important issue for fuel cells in cars or portable systems. The films are also chemically stable and can tolerate extreme fuel cell operating conditions like humidity and reactive gases. Although having promising advantages, manufacturing complexity and cost of CNT heating films prevent their wide application on a large scale. However, further research and developments in nanomaterial fabrication should improve scalability and affordability. Embedded CNT heating films are thus becoming a promising solution for the cold start performance of next-generation hydrogen fuel cells (Mohideen et al., 2020).

2.2.4 Comparative Efficiency and Response Time

The comparison of relative efficiency and response time is an important parameter to evaluate different heating methods employed in hydrogen fuel cells for cold start management. Efficiency will be the rate at which a heating technique changes input energy to useful heat; response time is the time it takes to reach the target temperature. Both parameters influence the start-up performance, energy consumption, and reliability of the fuel cell system in a low-temperature environment.

Resistive heating is one of the simplest and cheapest methods. It is moderately efficient with a relatively fast response time when the heating elements are near or within the MEA, respectively. The direct transformation of electric energy into heat allows accurate control; however, energy losses can be due to heat dissipation in the surrounding components. Its fast activation and simple integration notwithstanding, it applies to most fuel cell applications. Inductive heating has a better response time and localized efficiency than resistive methods. Because heat is produced inside the conductive parts by electromagnetic induction, it eliminates energy losses due to external dissipation. It is ideally suited for systems requiring fast and precise heating without physical contact but requiring more complex control systems and shielding.

Less common but very rapid and efficient is microwave heating when tuned to the properties of the target material. It is especially effective on dielectric or composite materials but requires sophisticated design for uniformity and safety. CNT heating films are very fast to respond and very thermally efficient at low power input. They disperse heat uniformly over the fuel cell layers to improve performance and cut start-up energy consumption. Overall, the best heating method depends on the fuel cell design, application, and performance goals (Duan et al., 2019).

2.3 DESIGN CONSIDERATIONS FOR HEATING CIRCUITS

2.3.1 Heat Transfer Requirements

Suitable heating circuits for hydrogen fuel cells must consider several technical parameters, such as cold start performance, energy efficiency, and system safety. Most important is the heat transfer requirement. A very good heating circuit ought to supply enough thermal power to take the electrolyte and some other vital elements to an optimum operating temperature of 60–80°C within a brief time. It involves accurate thermal modeling to calculate the required power according to the material mass, specific heat, and thermal conductivity.

Material selection is equally important. Heating elements should have good electrical resistance, good thermal conductivity, and chemical stability in a fuel cell environment. Common materials are metal alloys, carbon-based composites, and conductor polymers. All these materials must also be compatible with the structural and electrochemical parts of the fuel cell to avoid degradation or performance interference in the long term. Integration with the overall fuel cell architecture is another important design factor. Heating circuits must not be bulky or affect the fuel cell form factor. These should also be placed in a way that provides uniform heat distribution over the MEA without creating hot spots. Uniform heating keeps the membrane from thermal stress, mechanical deformation, and failure.

Electrical short circuits, overheating, or thermal damage to nearby components must be prevented by safety and insulation. This requires insulation layers, thermal barriers, and protective coatings. Also, sensors with feedback should be integrated to monitor temperature in real time and adjust power input dynamically. All these design features allow safe and effective heating with a fast start-up and a long service life of the fuel cell system, even under harsh environmental conditions (Pei et al., 2022).

2.3.2 Material Selection for Heating Elements

Material selection for heating elements of hydrogen fuel cells is a critical step in the design for heating system efficiency, durability, and safety. Since the function of these elements is to produce and transfer heat to an electrolyte and nearby components during cold starting conditions, the materials must possess the required electrical, thermal, and chemical properties. Among the requirements is high electrical resistivity so that the material can convert electrical energy to heat by Joule heating. Meanwhile, good thermal conductivity is required so that the heat generated is dispersed over the target area without localized hot spot occurrence.

Common materials are metal alloys like nichrome, stainless steel, which combine good resistivity, mechanical strength, and oxidation resistance. These can withstand repeated thermal cycles and keep their structural integrity in the chemically reactive and humid environment inside a fuel cell. In more advanced applications, carbon-based materials include CNTs and graphene. These offer excellent thermal and electrical properties, low weight, and high flexibility, and are therefore suitable for integration in compact and high-performance systems. Another important material selection is chemical stability. Heating elements are often near PEMs and/or other sensitive components, so materials must not be contaminated by fuel gases such as hydrogen and oxygen, nor degrade during operation. Heating materials must be compatible with the fuel cell MEA to prevent oxidation or interference with electrochemical reactions. Durability under thermal stress as well as mechanical loading is required. They must withstand rapid temperature changes without cracking or losing conductivity. The right materials ultimately ensure efficient heat generation, system longevity, and safe operation under varying temperature and load conditions.

2.3.3 Integration with Fuel Cell Architecture

Heating elements are an essential part of system design that allows effective thermal management without compromising performance, compactness, or safety of the fuel cell. Its aim during integration is to position the heating elements to deliver uniform and controlled heat to critical areas such as the electrolyte and MEA while remaining compatible with the overall structure and operation of the fuel cell stack. Proper integration avoids energy loss and thermal gradients and allows faster start-up under cold conditions.

One of the key considerations for integration is spatial configuration. Heaters must be incorporated in or placed near the membrane without interfering with the gas flow channels, current collectors, or sealing elements. Thin, flexible heating films are preferred, which conform to the cell layer shape and take up little space. In bipolar plate designs, resistive heaters may be directly incorporated in or behind the plates to heat adjacent layers. Electrical connectivity must also be controlled during integration. Electrical isolation of the heating circuit from electrochemical pathways is required to avoid short circuits or disturbances in current flow. Electrical segregation is maintained while heat transfer is achieved using insulation materials and thermal interface layers. Additionally, sensors and control electronics for monitoring temperature and power input should be integrated into the system. Mechanical compatibility is important, too. Heating elements must not delaminate or break due to thermal expansion, pressure change, or vibrations. The integration process should not make assembly and maintenance of the fuel cell stack difficult. By designing structural, electrical, and thermal compatibility, successfully integrated heating systems improve cold start performance while preserving fuel cell efficiency and life.

2.3.4 Safety and Insulation Aspects

The safety and insulation issues are of prime concern during the design and operation of heating circuits of hydrogen fuel cells under cold start conditions, where additional energy input to the system is required. Heating elements, if not properly managed, can cause electrical short circuits and thermal degradation of sensitive materials and fire hazards. Thus, thermal and electrical safety by appropriate insulation techniques is required for safe and reliable operation of the fuel cell. Among the safety issues is avoiding electric contact between heating elements and electrochemical components of the fuel cell. Since heating systems usually work at relatively high currents, unintended contact could cause short circuits or damage to the membrane and electrodes. To mitigate this, electrical insulation materials are used, such as ceramic coatings, flexible mica sheets, or high-temperature polymers. These give good dielectric separation while withstanding start-up operation heat.

Thermal insulation is also important so that heat is directed where it is needed – to the electrolyte assembly and MEA. Lacking proper insulation, heat may escape to surrounding structures, reducing system efficiency and producing uneven heating, which may stress the membrane or catalyst layers. Insulating materials include aerogels, foam insulators, and thermally resistant composites that contain and direct the heat flow. In addition, the system must have temperature sensors and thermal cut-off mechanisms to prevent overheating. These safety components monitor real-time conditions and shut down the heating circuit when temperatures become unsafe. This prevents damage and extends the life of the fuel cell system. Fire-resistant enclosures and tough circuit breakers are also used for automotive or industrial safety requirements. Overall, safety considerations and insulation contribute to a safe and stable operation of heating systems of hydrogen fuel cells.

2.4 ELECTRICAL CONTROL AND POWER MANAGEMENT

Electrical control and power management are important heating elements in hydrogen fuel cells, especially during cold start operations. Heating circuit efficiency and safety are directly related to the regulation, distribution, and monitoring of electrical input. A good control system ensures that heating elements are given just enough power to heat the electrolyte without overheating, excessive energy use, or damage to fuel cell components. Voltage regulation and current regulation are among the basic elements of electrical control. Since fuel cells work at lower voltages, heating systems will often need an external power source or boost converter to obtain the power. Power electronics are employed to manage this power – PWM controllers, DC converters. Such devices allow temperature feedback-controlled fine-tuning of heating output by changing the duty cycle or current levels in real time based on temperature feedback.

Thermal sensors like thermocouples and resistance temperature detectors are integrated into heating circuits for continuous monitoring of critical areas such as MEAs. This data is fed into a control unit that dynamically adjusts power to the heating elements to keep target temperatures without overshooting. In advanced systems, this is automated using microcontrollers, embedded processors with safety protocols, diagnostic features, and predictive algorithms.

Power management also involves load balancing, especially when heating is required in combination with other electrical loads like compressors, pumps, or vehicle electronics. Intelligent energy management prevents heating processes from drawing excessive power, especially in battery-assisted or hybrid fuel cell systems. Good electrical control and power management are thus required for fast, safe, and efficient cold start operations of hydrogen fuel cells.

2.4.1 Power Source Compatibility (DC/AC)

In heating circuits for hydrogen fuel cells, power source compatibility is very important, especially for cold start problems. Heating elements need a steady and matched power supply. Direct current (DC) versus alternating current is a decision based on the application, system configuration, and available energy sources. Most hydrogen fuel cells, especially automotive and portable ones, operate on DC, so heating circuits are usually DC compatible. It enables a simplified system integration without adding additional power conversion equipment. DC-compatible heating systems can be started directly from fuel cell output or with an auxiliary battery at start-up. This arrangement reduces conversion losses and allows precise voltage and current control by simple electronic devices such as PWM units or DC-DC converters. These controllers adjust heating power based on temperature sensor feedback for safe and effective thermal management.

In stationary or industrial fuel cells where AC power is available from the grid, heating circuits may be designed for AC compatibility instead. However, if the fuel cell works on DC, then transformers and rectifiers are required. AC systems can handle higher voltages and power levels, but add complexity in insulation, control systems, and safety, especially during DC conversion for internal components. A hybrid configuration using both AC and DC sources often uses dual compatible heating elements and converters for greater flexibility and reliability. Such systems generally include automatic switching with voltage regulation for easy power transitions. In general, power source compatibility ensures heating efficiency optimization as well as safety, responsiveness, and operation reliability of the hydrogen fuel cell system under diverse conditions.

2.4.2 Voltage Regulation and Current Regulation

Regulating voltage and current is an essential factor for safe and effective heating circuit operation in hydrogen fuel cell systems, especially at cold start. Heater elements need exact amounts of electrical energy to produce heat without damaging fuel cell components or wasting power. Controlling both voltage and current assures stable performance, avoids overheating, and optimizes energy use in the heating phase. Voltage regulation provides a constant voltage supply to heating elements even when input power changes. Such fluctuations are normal at start-up as load conditions vary rapidly, and output may not be stable from fuel cells or batteries. To counter this, voltage regulators or DC-DC converters step up or down the voltage as required. This stability is required for uniform heat generation over the MEA.

Current regulation is equally important since the heat output of the resistive element is proportional to the current passing through it. Excess current can overheat, damage sensitive components, and shorten system life. Current controllers/limiters – sometimes guided by real-time temperature feedback – adjust current flow dynamically to keep target temperatures without energy wasting or safety compromise. Most modern fuel cell designs combine voltage regulation with current regulation in integrated control units. These units analyze sensor data and tune power delivery with high accuracy using microcontrollers/digital signal processors. Voltage regulation and current regulation ensure safe and effective operation of heating circuits even in extreme cold conditions.

2.4.3 Thermal Sensors and Feedback Loops

Thermal sensors, as well as feedback loops, are of prime importance for heating circuit design and operation in hydrogen fuel cells, especially in cold start conditions. They provide real-time temperature monitoring and control over major components, especially the PEM, for safe and efficient heating. Without accurate sensing and responsive regulation, heating circuits may fail to perform properly and delay start-up, or overheating may damage sensitive materials and shorten fuel cell life. Thermal sensors like thermocouples, resistance temperature detectors, and thermistors are placed near heating elements and MEAs, respectively. These sensors send continuous temperature readings to a control unit. Sensor selection is based on required accuracy, response time, operating temperature range, and environmental conditions in the fuel cell stack. For example, RTDs are used for accuracy and long-term stability, whereas thermocouples are used for high-temperature environments.

The feedback loop compares the sensor data with temperature setpoints defined in the sensor data. Abnormality leads to adjustment of the power supply to heating elements, thereby restoring temperature targets. This closed loop avoids overheating, improves energy efficiency, and maintains uniform heat dispersion. Advanced implementations may use proportional-integral-derivative (PID) controllers or AI algorithms to forecast temperature trends and adjust heating inputs accordingly. Feedback systems are also required for safety, besides efficiency. Unusual temperature rises may cause alarms, or the heating circuit may be shut down. Thermal sensors and feedback loops allow hydrogen fuel cells to carry out cold start operation with high durability and stability.

2.4.4 Load Management during Start-Up

During start-up, load management is very important in hydrogen fuel cell operation, particularly cold start, when more power is required to heat the electrolyte and get the system to an optimal temperature. In this phase, many components, including heating circuits, sensors/control units, and auxiliary devices like pumps or compressors, may be using power simultaneously. This sudden increase in electrical load without coordination may overload the power supply, lead to voltage drop, degrade efficiency, or cause system faults. Good load management distributes available power so that critical functions like electrolyte-heating are prioritized over secondary tasks. This is important in fuel cell vehicles and portable systems where start-up power is often from a small battery or capacitor bank. Therefore, smart power controllers and energy management units may sequence subsystem activation according to priority and real-time operation conditions. In staged allocation strategies, for example, the heating circuit is first powered to the target temperature, and then other components are gradually engaged.

Other load-shedding techniques include temporarily shutting down non-essential components until the fuel cell starts producing its own power. Real-time current and voltage monitoring allows dynamic load priority adjustment while preserving system stability. Advanced designs include predictive algorithms/control software for predicting power demand trends. Managing electrical loads during start-up can ensure reliable ignition, shorten warm-up time, avoid electrical stress on components, and safe transition to steady state operation in various environmental and operating conditions.

2.5 THERMAL MANAGEMENT AND DISTRIBUTION

Thermal management and distribution are required to achieve good temperature stability in hydrogen fuel cells during cold start and steady-state operation. Good thermal control ensures that the electrolyte, electrodes, and catalyst layers work efficiently without thermal gradients causing mechanical stress, performance loss, or long-term degradation. Localized heating must raise the electrolyte to a conductive temperature without destroying nearby materials during start-up, so heat distribution is critical.

Temporally uniform MEA is the main goal. Uneven heating can produce hot spots or cold zones of differential expansion, higher resistance, or membrane cracking. This is prevented by positioning heating elements correctly and designing conductive paths so that heat travels more evenly. Graphite plates or metallic layers are high-thermal-conductivity materials used for heat transfer across the cell. External insulation improves efficiency further by minimizing heat loss to the environment, particularly in low-temperature conditions. Thermal interface materials (TIMs) are placed between layers to improve contact, thermal resistance, and heat flow in the stack. Advanced designs have thermal sensors that measure temperature distribution dynamically to avoid overheating.

Waste heat from fuel cells or auxiliary subsystems can be recovered and diverted to pre-heat the stack in automotive and stationary systems. Comprehensive thermal management and distribution strategies for hydrogen fuel cells are able to realize fast start-up time, longer service life, and good performance under different environmental and load conditions.

2.5.1 Uniform Heat Distribution Techniques

Uniform heat distribution in hydrogen fuel cell systems is required so that all parts, especially the MEA, can reach and maintain their optimal operating temperatures. Uneven heating during cold starts can result in performance losses, mechanical stress, and long-term degradation. This uniform distribution is possible only if the heating elements are designed well and thermally conductive materials are integrated into the cell architecture. Typical techniques include using distributed resistive heating elements patterned or layered over the fuel cell's surface area. This arrangement provides consistent heat without localized overheating or cold spots. Flexible heating mats and thin-film heaters embedded in bipolar plates enable targeted, even heating without significant bulk or design complexity.

Thermal uniformity may also be improved by incorporating high-conductivity materials – graphite, aluminum, or copper alloys – into cell plates or interlayers to promote lateral heat spreading. Their placement provides efficient heat conduction pathways and reduces temperature gradients. Also, optimized flow channel designs for reactant gases and coolants may allow balanced heat exchange over the active area to promote uniformity. Advancement techniques use TIMs to lower contact resistance between layers for better heat transfer. Combined with real-time thermal monitoring and smart control systems, these setups can dynamically adjust heating patterns to keep the operating temperature constant.

2.5.2 Mitigating Thermal Stress and Gradients

Thermal stress and temperature gradient minimization in hydrogen fuel cells is an important issue for long-term durability and steady performance during start-up and in fluctuating operating conditions. Thermal stress occurs when different regions of a fuel cell heat or cool differently, so materials expand and contract differently. This mismatch can cause mechanical deformation, layer delamination, membrane cracking, and failure of sealing components. Temperature gradients become particularly difficult during cold starts, when heating systems quickly warm the electrolyte and internal parts. To mitigate these effects, a gradual, uniform temperature rise along the cell stack is necessary. One such approach involves uniformly spaced heating elements supplying consistent heat to the MEA. Such systems are often controlled by real-time temperature monitoring/feedback control loops, adjusting power levels to avoid localized overheating. Pre-heating methods that gradually raise the system temperature before full operation can also prevent sharp thermal gradients and associated mechanical stress.

A second factor reducing thermal stress is material engineering. Picking elements having matched thermal expansion coefficients ensures that all materials expand and contract at the same rate, thus minimizing mechanical failure risk and enhancing structural integrity. Flexible sealing materials and reinforced membranes are added for thermal cycling resistance. Heat-spreading materials like graphite or metallic layers are also used to disperse heat laterally, such that they do not produce heat or cold spots. Also, effective thermal insulation and optimized flow field designs maintain uniform temperature distribution.

2.5.3 Role of Phase Change Materials

Phase change materials (PCMs) are an important part of the thermal management of the hydrogen fuel cells, especially in cold start problems and at the optimal operating temperatures. PCMs absorb, store, and release large amounts of thermal energy during phase transitions – typically from solid to liquid and vice versa. This makes them good thermal buffers that maintain constant temperature in the fuel cell and protect sensitive components from sudden changes in temperature. In cold starts, PCMs gradually lose stored heat to warm electrolyte and other surrounding components to permit ionic conductivity within the membrane without reliance on outside heating sources. In reverse, under normal operation, if internal temperatures rise too quickly, PCMs absorb the excess heat as they melt, preventing overheating. This dual capability improves start-up reliability and steady-state thermal regulation.

Placement of PCMs in the fuel cell system is engineered. They are placed in or around the cell housing, near heating circuits, or in thermal interface zones where rapid temperature changes are likely. Integration should avoid interference with gas flow paths or electrical insulation. The PCMs and volumetric expansion control during phase transitions are usually encapsulated to achieve long-term stability and performance. PCM selection is influenced by phase transition temperature, thermal conductivity, and system-operating environment compatibility. Organic materials like paraffin and inorganic salts have good thermal behavior and are often adjustable.

2.5.4 Thermal Interface Materials (TIMs) in Stack Design

TIMs are an important pathway for heat transfer between adjacent layers and components in hydrogen fuel cell stack design. In a typical stack consisting of bipolar plates, membranes, catalysts, and structural supports, even after precise manufacturing, micro-air gaps and surface irregularities may remain despite exact manufacturing. These gaps inhibit thermal conduction, and TIMs are used to fill them, reducing thermal resistance and enhancing heat flow across interfaces.

A uniform temperature distribution within the fuel cell stack is critical for performance, durability, and safety during cold starts and normal operation. These TIMs help in this by distributing heat produced by electrochemical reactions or supplied by heating circuits uniformly, minimizing temperature gradients, and preventing localized hot or cold spots. This is especially important when using embedded heating elements, as uneven contact leads to thermal stress. TIM selection is based on thermal conductivity, mechanical compliance/electrical insulation, and chemical compatibility with fuel cell materials. Common options are thermal greases and PCMs, elastomer pads, and graphite-based sheets. In hydrogen fuel cells, electrically non-conductive TIMs are preferred to avoid short circuits while preserving high thermal performance. Also, TIMs must tolerate repeated thermal cycling without degradation to maintain long-term stability under frequent start–stop cycles and at different ambient conditions.

2.6 COLD START PERFORMANCE OPTIMIZATION

Cold start performance optimization of hydrogen fuel cells is an important design issue for automotive and portable applications working in sub-zero conditions. Low-temperature PEMs and electrolytes show poor ionic conductivity, leading to poor electrochemical reactions and poor overall efficiency. To meet these challenges, engineers employ thermal, electrical, and material-based solutions to bring the system up to operating temperature with minimum energy consumption and mechanical stress. A core strategy is to integrate fast and effective heating systems like resistive heaters, inductive elements, and nanomaterial-based films, to warm electrolyte and catalyst layers. These heating elements are often incorporated in or around the MEA to deliver the targeted thermal energy where it is needed. Real-time temperature sensors and feedback control loops regulate heating rates to avoid thermal shock and maintain uniform temperature distribution.

Material optimization is equally important. Good performance is achieved by low-temperature-tolerant membranes, good cold start catalysts, and freeze-resistant components. Supporting materials such as PCMs and TIMs improve heat retention and distribution further. These materials store thermal energy in the storage during normal operation and release it during start-up to shorten warm-up and reduce energy demand. Intelligent control algorithms also manage power allocation, heating sequences, and load balancing on start-up. Such systems may direct battery or external power first to critical components for faster activation at lower energy cost.

2.6.1 Heating Time versus Efficiency Trade-Off

Heating time/efficiency balance is a fundamental problem for cold start performance optimization of hydrogen fuel cells. Low-temperature electrolyte and its constituents must be quickly warmed to recover ionic conductivity to permit electrochemical reactions. Heating rate influences energy efficiency and system durability. Fast heating reduces start-up delays and allows immediate power generation, an advantage for automotive and emergency applications, but it usually requires high power input from onboard batteries or external sources, a drawback that reduces efficiency and adds load on auxiliary systems. Instead, a slow, more controlled heating process saves energy and reduces thermal gradients that may induce mechanical stress, lamination, or damage of the MEA. Such an approach improves fuel cell longevity and reliability but increases start-up time, which is unacceptable in applications that need a fast response.

Rapid versus gradual heating selection depends on system priority – immediate performance/long-term efficiency. To fill this gap, advanced control strategies are applied. Adaptive predictive algorithms adjust power delivery to heating elements based on ambient conditions, prior start-up patterns, and real-time temperature feedback. Hybrid methods are also possible: rapid heating to a threshold temperature is followed by a low-power phase to reach full operational readiness without excessive energy consumption. Besides, energy recovery systems and materials with high thermal retention contribute to this balance.

2.6.2 Temperature Ramp-Up Profiles

Temperature ramp-up profiles for hydrogen fuel cells define a controlled temperature rise from the cold state to the optimal operating range at start-up. These profiles are required so that the fuel cell can reach functional temperatures safely and without thermal stress in internal components. Designing a good ramp-up profile requires determining the heat application rate and uniformity to the electrolyte as well as other stack components. The aim is to reduce start-up time while avoiding sharp thermal gradients, which may damage sensitive layers such as PEM or catalytic-coated electrodes.

Sometimes in cold start scenarios, ramp-up profiles are executed separately. The initial phase is usually faster, and the system is above the critical threshold temperature for ionic conducting. Following this is a slower, more controlled heating phase to full operating temperature with even heat distribution to the stack. Extremely rapid heating in localized areas may cause uneven expansion and membrane cracking or seal failure. On the other hand, too slow ramp-up delays start-up, increases auxiliary energy consumption, and lowers system efficiency. To keep the intended profile, control systems with thermal sensors monitor temperature progression and adjust heating input in real time. Based on component materials, ambient conditions, and application requirements, predefined ramp-up curves are drawn. Advanced systems may modify these curves dynamically based on predictive algorithms and historical performance traces.

2.6.3 Impact on Fuel Cell Longevity

Cool start strategies, along with associated thermal management, influence fuel cell longevity and the performance of hydrogen fuel cell systems. Repeated sub-zero temperature exposure and mechanical and chemical stresses during start-up may accelerate degradation of key components, including the membrane, catalyst layers, gas diffusion layers, and seals. Probably the most important is ice formation in the cell during shutdown and start-up, which may physically damage the MEA due to expansion and contraction forces. If not properly tempered with appropriate heating and water management strategies, this may cause irreversible performance loss in the long run.

Aggressive heating methods that aim to rapidly warm up the electrolyte might allow faster start-up but may produce thermal gradients that cause delamination, microcracks, and seal failure. These localized damages accumulate during repeated cold start cycles and affect the mechanical integrity as well as the electrochemical efficiency of the stack. Non-uniform heating may also create hot spots that damage the catalyst layers or cause oxidation reactions in unintended places that shorten the fuel cell's life. To counter these risks, fuel cell designers adopt intelligent heating controls, thermal buffers like PCMs, and advanced TIMs for uniform heat distribution and thermal stress reduction. Optimized ramp-up profiles and controlled start-up sequences balance start-up speed trade-off with component wear. Also, selecting materials with high freeze–thaw durability and incorporating anti-freeze additives in the water management system can improve resistance to repeated cold cycles. Monitoring systems also record degradation indicators over time to enable predictive maintenance and system adjustments before critical failures. Considering such factors during design and operation can significantly extend fuel cell life while maintaining consistent performance and reliability even during frequent or extreme cold start conditions.

2.6.4 Real-Time Diagnostics and Monitoring

Diagnostics and monitoring in real time are extremely important for cold start performance optimization and long-term reliability of hydrogen fuel cells. In their cold start phase, fuel cells are particularly sensitive to thermal, mechanical, and electrochemical stresses. These conditions are detected and responded to immediately by real-time monitoring systems, allowing better control of heating strategies and avoiding failures. Sensing devices embedded in the fuel cell stack and associated components collect temperature, voltage, current, humidity, and pressure data for operators or control units to make informed decisions during the critical start-up period.

One of the major tasks of real-time diagnostics is temperature gradient monitoring across the fuel cell for uniform heating. Incorrectly heated components may produce thermal stress, membrane deformation, or ice formation, compromising the structural integrity and performance of the system. Through feedback loops of thermal sensors, the heating elements can vary power output dynamically to keep a temperature ramp-up profile controlled. Ionic conductivity and voltage monitoring across individual cells, in addition to temperature detection, reveal anomalies like poor contact, material degradation, or localized flooding. Advanced systems integrate predictive analytics and machine learning algorithms, which use historical, real-time inputs to forecast possible failures and inefficiencies. This enables proactive maintenance and adaptation of start-up protocols depending on the environment/system's health. Furthermore, real-time diagnostics enable data logging for performance tracking, regulatory compliance, and warranty validation. In automotive or remote applications, remote monitoring capabilities are especially useful to perform system diagnostics without physical inspection. Overstress thresholds may be alerted, and automated shutdown mechanisms may be activated to prevent irreversible damage to the system. Overall, real-time diagnostics/monitoring provides the basis for intelligence and safety to enable robust, efficient, and repeatable cold start performance of hydrogen fuel cells.

2.7 ADVANCED MATERIALS AND TECHNOLOGIES

Modern advanced materials and technologies are central to cold start performance and overall efficiency of hydrogen fuel cells. Because the sub-zero operation of such systems requires high durability, ionic conductivity, and thermal resilience requirements, conventional materials are not suitable for this application. To overcome these limitations, researchers and engineers have developed special materials for components like the membrane, catalyst gas diffusion layers, and thermal management systems. A major step forward is the fabrication of low-temperature-tolerant PEMs retaining good ionic conductivity at freezing temperature. These membranes use less energy for heating and permit quicker start-up times without losing performance.

Parallel to this, new catalyst formulations containing platinum alloys/nano-structured supports are presented to maintain catalytic activation at lower temperatures. These catalysts increase electrochemical reaction rates at cold starts and resist degradation under thermal cycling. Likewise, gas diffusion layers of advanced carbon materials or composite structures improve water management and frost resistance, thus reducing the risk of ice formation damaging the cell. TIMs are progressively combined with PCMs to achieve uniform heat distribution and energy storage in operation. They capture and reuse waste heat for future start-ups. Another technological innovation is the embedded heating elements made of CNT films or metal foils, which heat the MEA very fast and evenly. Smart materials and nanotechnology also provide sensing and actuation functions, allowing real-time adaptation to start-up conditions. To meet the harsh climates and mobile applications, fuel cells are increasingly used, so the combination of these advanced materials and technologies is mandatory for long-term, reliable operation. They help overcome the cold start challenge and promote wider adoption of hydrogen energy systems.

2.7.1 Conductive Polymers and Composite Materials

Conductive polymers and composite materials are now critical components for cold start capability and overall efficiency of hydrogen fuel cells. Traditional materials used for fuel cell construction have thermal and electrical conductivity limitations, flexibility limitations, and mechanical stress limitations under sub-zero conditions. Conductive polymers like polyaniline, polypyrene provide efficient flexibility, conductivity, and environmental stability for use in MEA and heating elements. These materials may be engineered to maintain low temperature performance while maintaining constant ionic and electrical conductivity when starting the fuel cell in a cold environment. Composite materials containing conductive polymers and CNTs, graphene, and metal nanoparticles add functional properties. Their composites show improved thermal stability, electrical conductivity, and mechanical strength critical for repeated freeze–thaw cycles. When fabricated in heating circuits or incorporated into the membrane support structure, they ensure uniform heat transfer, shorten start-up time, and avoid localized overheating/cold spots that damage cell components. Composite materials also allow light and flexible designs – useful in automotive and portable fuel cell applications. Their structural adaptability allows better integration in compact systems without performance compromise. Also, these materials allow self-regulating thermal responses as conductivity changes with temperature, providing another layer of passive control during start-up.

Additional developments in material science lead to enhanced characteristics of conductive polymers and composites for much better toughness, lower price, and manufacturability. Utilizing these materials, engineers may develop fuel cells that can start faster in cold conditions while retaining good performance and durability, which may accelerate hydrogen energy technology adoption in real-world applications.

2.7.2 Thin-Film Microheaters and MEMS Integration

Thin-film microheaters and micro-electromechanical systems integration are two advanced technological solutions for cold start performance of hydrogen fuel cells. Thin-film microheaters are miniature heating elements deposited by precise deposition of conductive materials like platinum, nickel, or polysilicon on glass, ceramic, or polymer films. These heaters have very fast thermal response times and accurate temperature control, so they are suitable for fast and localized heating applications such as warming of electrolyte/catalyst layers in the initial cold start. Because microheaters have low thermal mass, they consume little energy while providing targeted heat, supporting energy-efficient start-up operations. MEMS completes microheaters by allowing miniature sensors, actuators, and control systems to be integrated directly in fuel cell architecture. They may be capable of monitoring important parameters as temperature, humidity, pressure, and gas concentration, in real time through control algorithms that dynamically adjust heating profiles and power distribution. The small size of MEMS devices allows their embedding within or adjacent to MEA without adding bulk or weight, which is very useful in automotive and portable fuel cell applications.

Thin-film microheaters in combination with MEMS allow localized on-demand heating of regions of the fuel cell more susceptible to cold-induced performance drops. This selective heating approach saves energy usage and thermal gradients, which can induce mechanical stress or uneven reaction rates. In addition to this, the scalability and manufacturability of MEMS-integrated systems make them suitable for mass production and provide a route to low-cost, high-performance fuel cells. They allow a smarter and more resilient cold start strategy to achieve fast activation, protect sensitive materials, and extend system lifespan in variable environmental conditions. As such, they are fundamental parts of next-generation fuel cell designs.

2.7.3 Self-Regulating Heating Materials

Self-regulating heating materials are an effective and safe way to manage thermal conditions during the cold start of hydrogen fuel cells. These materials can adjust the heat output according to the ambient temperature with no complex external controls. They are usually made of polymers containing conductive particles like carbon black/metal oxides and have positive temperature coefficients (PTCs). At higher temperatures, their electrical resistance increases, reducing the current flow and, consequently, the heat generation. Its inherent self-limiting feature prevents overheating, and uniform heating is maintained across the fuel cell components.

In cold start scenarios, self-regulating heating materials are especially useful in delivering targeted heat where and when required. If applied to areas like the electrolyte, gas diffusion layer, or interfaces of the MEA, these materials heat up upon power application. As the local environment temperature increases to the desired operating range, the materials reduce thermal output to avoid hot spots and thermal stress. With this dynamic response, energy efficiency is improved, and sensitive fuel cell components are not damaged by excessive heat. The flexible and lightweight nature of self-regulating heating films allows easy integration into various fuel cell architectures and long-lasting, curved, and compact geometries found in automotive applications. Thermostats, sensors, and external control circuitry for heat management are also eliminated in system design. Besides, they are reliable and long-lasting, ideal for repeated freeze–thaw cycles. In cold climates and off-grid environments, self-regulating heating materials are of prime importance to ensure consistent performance, quick start-up, and system longevity while minimizing energy consumption and design complexity.

2.7.4 Nanotechnology for Targeted Heat Generation

Nanotechnology enables targeted heat generation to enhance cold start solutions of hydrogen fuel cells. At the nanoscale, materials possess special thermal, electrical, and structural properties that may be exploited for heating process precision, efficiency, and speed in fuel cell systems. Nanomaterials like CNTs, graphene, metal nanoparticles, and nanowires are excellent thermal and electric conductors, permitting localized heat production with minimum power input. Such materials may be integrated directly into the MEA or spread over other critical areas like gas diffusion layers or bipolar plates.

Another benefit of using nanotechnology in cold start systems is that heat generation is focused exactly where it is required – around the electrolyte or catalyst layer. Hence, this targeted approach reduces power consumption and start-up time while avoiding excessive heat exposure of non-essential components. Nanomaterials may be embedded in films or coatings to form nanoscale heaters that respond to power immediately. Such nanoscale heaters disperse heat uniformly and react quickly to environmental changes, making them suitable for dynamic cold start conditions. Nanostructured materials also allow multifunctional components combining heating, sensing, and structural support. CNT networks, for example, can be both heaters and sensors that give feedback on temperature distribution in the cell. This dual functionality enables smarter fuel cell systems to adjust the heating profile according to live data. Also, nanotechnology allows ultrathin, flexible parts to be formed to fit complex geometries with no performance compromises. As fuel cells move into colder climates and mobile platforms, nanotechnology for targeted heat generation will become critical. This opens a route to safer and faster cold starts, supporting a wider application of hydrogen energy systems.

2.8 FUTURE TRENDS AND RESEARCH DIRECTIONS

Future trends and research directions of electrolyte-heating circuits of hydrogen fuel cells are increasingly concerned with increasing the efficiency, reliability, and adaptability in response to diverse environmental conditions. One of the main areas of development is the intelligent thermal management systems, combining sensors, real-time monitoring, and adaptive control algorithms for heating performance optimization during cold starts. Such systems aim at minimizing energy consumption while ensuring uniform electrolyte-activation speed in extreme climates. It is also working on smaller and lighter heating elements for portable and automotive applications, where space and energy are limited.

Material innovation drives further progress with research focusing on novel composites and self-heating materials. Hence, nanostructured coatings with excellent thermal conductivity, faster response times, and better durability. Eventually, materials will combine heating capability with structural strength and chemical resistance. All these advancements may allow heating systems to be installed in fuel cell stacks without external circuits/bulky insulations. In addition, scalable manufacturing methods for microheaters as well as MEMS devices are investigated for mass production and low cost. A further major direction is the hybrid systems combining passive and active heating strategies, such as waste heat recovery/PCMs, together with embedded microheaters. Such an approach optimizes energy use and maintains operational stability during start-up and continuous operation. Further research involves artificial intelligence and machine learning for system behavior prediction and thermal profile adjustment on demand based on environmental and operational data. As hydrogen fuel cells take off in transport, stationary power, and aerospace applications, cold start solutions will only get stronger. Continued interdisciplinary research will be needed to design systems that are efficient but also flexible, sustainable, and scalable.

2.8.1 Smart Heating Circuits with AI Control

The smart heating circuits with AI control are an important step toward cold start performance improvement of hydrogen fuel cells. Such systems employ artificial intelligence to dynamically manage and optimize the thermal behavior of heating elements during cold start-up. Typical heating systems use fixed parameters or manual control, which may lead to inefficient energy use or uneven heating. Rather, AI-integrated heat circuits process real-time data from several sensors embedded in the fuel cell, such as temperature, voltage, humidity, and current flow, to adjust heating parameters precisely and quickly.

Smart heating circuits learn from past start-up habits and environmental conditions to select the best heating profile for future starts via machine learning algorithms.

This allows adaptive thermal management where heat delivery to other components is reduced selectively to those areas that require more heat, such as electrolyte/catalyst zones. Besides, this precision reduces energy consumption and the risk of thermal stress, degradation, or uneven electrolyte activation. Moreover, AI systems can monitor the health and performance of the heating elements over time – detecting wear or malfunctions early and sending out alerts for preventive maintenance. Load balancing and power optimization based on fuel cell system state and external power sources are also supported by smart heating circuits. For instance, during vehicle ignition, the AI may prioritize power flow between the heating system and critical components for a stable and fast start-up. These systems are particularly useful in variable climate conditions where external temperatures and humidity vary wildly. AI helps to improve cold start reliability as well as the safety and longevity of fuel cell systems. AI technologies are expected to become more central and sophisticated in smart heating circuit design.

2.8.2 Integration with Vehicle Thermal Systems

The combination of electrolyte-heating circuits and vehicle thermal systems is a promising strategy for cold start performance enhancement and energy efficiency improvement of hydrogen fuel cell vehicles. In modern vehicles, advanced thermal management systems control engine temperature, cabin heating, battery conditioning, and component cooling. Integrating the fuel cell heating requirements into this thermal framework enables the sharing of resources like coolant loops, heat exchangers, and thermal sensors for temperature control optimization without duplication. In cold start scenarios, the integrated system can allocate heat transfer to fuel cell stack critical components, such as electrolyte assembly and MEA, for instance. Waste heat from other vehicle systems, like electric motors/battery modules, can be directed toward pre-heating the fuel cells. This lowers the electrical load of resistive/embedded heating elements and reduces warm-up times. Coordinated thermal flow control ensures heat is distributed efficiently among different systems according to real-time demand, improving start-up reliability and energy utilization.

Thermal integration also permits centralized controllers and feedback loops to monitor vehicle temperatures and dynamically adjust heating strategies. With this synergy between vehicle systems, the overall thermal architecture is simplified, and redundant components are eliminated, thereby saving weight, cost, and complexity. Moreover, by using shared sensors and control electronics, manufacturers can improve system diagnostics, predictive maintenance, and long-term reliability. In cold weather/sub-zero environments integrated thermal systems can maintain the electrolyte temperatures at the same time as supporting passenger comfort and battery performance. The standardized operation of the hydrogen fuel cell system in combination with vehicle thermal networks will play an important role as fuel cell vehicles become commercially viable.

2.8.3 Energy Harvesting for Pre-Heating

Energy harvesting for pre-heating of hydrogen fuel cells is proposed to meet the energy demands of cold starts. The electrolyte and other critical components should be pre-heated to obtain good ionic conductivity, and performance degradation at low temperature should not be impaired. Typical heating methods use external power sources, which may not always be effective and available, especially in off-grid/mobile applications. Alternatives include energy harvesting, where ambient or residual energy from the environment or vehicle operation is captured and used to initiate heating.

One type of energy harvesting is thermoelectric generators that change waste heat from vehicle components like exhaust systems, battery packs, or motors to usable electrical energy. This harvested energy can be stored in capacitors or batteries and used to heat elements during the start-up phase. One method involves regenerative braking systems in electric or hybrid vehicles, collecting kinetic energy to be used for pre-heating the fuel cell stack. Also, photovoltaic cells embedded in the vehicle exterior or surface panels may provide a continuous and renewable low-level power source for sustaining baseline temperatures or starting slow pre-heater cycles. Using energy harvesting technologies reduces the dependence on the main battery or fuel supply during start-up to save energy for propulsion and improve system efficiency. In addition, they contribute to sustainability targets by utilizing otherwise wasted energy and emission reductions due to cold start inefficiencies. Modern control systems can monitor harvested energy status and allocate it to heating when needed most. Future developments in materials science and electronics should make energy harvesting components smaller, more responsive, and suitable for integration in fuel cell vehicle architectures. As such technologies mature, they will be critical to an autonomous and efficient cold start strategy.

2.8.4 Low-Cost, Scalable Heating Solutions for Commercial Fuel Cells

Low-cost, scalable heating solutions are required for a wide commercial deployment of hydrogen fuel cells in automotive, residential, and portable power applications. High-performance materials or complex integration techniques are typical heating technologies used in cold start scenarios. Heating systems must be designed for performance as well as for affordability, simplicity, and ease of production on a large scale for commercial viability. One such balance is achieved using economically viable materials such as printed carbon-based heating films, conductive inks, and polymer composites that are mass-produced by roll-to-roll or screen printing.

These are thermally conductive and can be directly incorporated into fuel cell components like bipolar plates or electrolyte membranes without the need for bulky and expensive external heaters. Flexible substrates permit fitting these heating elements into complex geometries for better thermal contact and heat transfer efficiency. Simple resistive circuits also may be fabricated from cheap metallic wires or foils and operate with little power electronics. Modular heating units can be standardized to fit various fuel cell sizes and configurations, reducing both design and maintenance steps. Systems complexity is also being reduced through passive or hybrid heating approaches, which combine low-cost electric heating with appropriate insulation, waste heat, and PCMs. Such a setup may keep target temperatures at low active energy inputs. In addition, open-source control systems and cheap microcontrollers are adapted to basic thermal management without expensive proprietary software/hardware. These scalable heating solutions prioritize affordability and manufacturability to allow fuel cell systems to be deployed in more markets and accelerate the clean hydrogen energy transition in developed as well as emerging economies.

2.9 CONCLUSIONS

Cold start capability is an important performance metric for hydrogen fuel cells in cars, portable devices, and cold environments. In this chapter, low-temperature challenges were reviewed for PEM electrolytes. These include lower ionic conductivity, water freezing, and mechanical stress. The report also covered electrolyte-heating solutions, including resistive, inductive, microwave, and CNT-based systems. The chapter evaluated their efficiency, response times, and integration methods. Heat transfer needs, material choice, integration in fuel cell setup, and strong safety considerations are key design considerations for consistent and reliable heating. Such control techniques as voltage regulation and current regulation, thermal sensing, and smart load management are required for quick start-up without compromising energy efficiency or component lifespan. Other thermal management techniques, like uniform heat distribution methods, PCMs, and TIMs, improve temperature stability and reduce thermal stress. Optimization strategies like customized heating time-efficiency balances, clear temperature ramp-up profiles, and predictive control algorithms allow faster activation while preserving durability. Advanced materials, MEMS-based microheaters, self-regulating heating films, and nanotechnology open up new opportunities for precise, targeted, and efficient heating. AI-driven smart heating circuits, vehicle-wide thermal systems integration, pre-heating energy harvesting, and low-cost, scalable designs are futuristic trends supporting a wider application of hydrogen fuel cell systems.

REFERENCES

Duan, C., Kee, R., Zhu, H., Sullivan, N., Zhu, L., Bian, L., Jennings, D., & O'Hayre, R. (2019). Highly efficient reversible protonic ceramic electrochemical cells for power generation and fuel production. *Nature Energy*, 4, 230–240. https://doi.org/10.1038/S41560-019-0333-2

Guo, S., Jiang, L., Li, Y., Zhong, P., Ismail, S. A., Norby, T., & Han, D. (2024). From electrolyte and electrode materials to large-area protonic ceramic fuel cells: A review. *Advanced Functional Materials*, 34. https://doi.org/10.1002/adfm.202304729

Lu, N., Zhang, Z., Wang, Y., Li, H., Qiao, S., Zhao, B., He, Q., Lu, S., Li, C., Wu, Y.-M., Zhu, M., Lyu, X., Chen, X., Li, Z., Wang, M., Zhang, J, Tsang, S.-C., Guo, J., Yang, S., Zhang, J., Deng, K., Zhang, D., Ma, J., Ren, J., Wu, Y., Zhu, J., Zhou, S., Tokura, Y., Nan, C., Wu, J., & Yu, P. D. (2022). Enhanced low-temperature proton conductivity in hydrogen-intercalated brownmillerite oxide. *Nature Energy*, 7, 1208–1216. https://doi.org/10.1038/s41560-022-01166-8

Mohideen, M. M., Liu, Y., & Ramakrishna, S. (2020). Recent progress of carbon dots and carbon nanotubes applied in oxygen reduction reaction of fuel cell for transportation. *Applied Energy*, 257, 114027. https://doi.org/10.1016/j.apenergy.2019.114027

Nguyen, H., & Shabani, B. (2021). Review of metal hydride hydrogen storage thermal management for use in the fuel cell systems. *International Journal of Hydrogen Energy*, 46, 31699–31726. https://doi.org/10.1016/J.IJHYDENE.2021.07.057

Pei, P., Fu, X., Zhu, Z., Ren, P., & Chen, D. (2022). Activation of polymer electrolyte membrane fuel cells: Mechanisms, procedures, and evaluation. *International Journal of Hydrogen Energy*. https://doi.org/10.1016/j.ijhydene.2022.05.228

Sharifvaghefi, S., & Zheng, Y. (2022). Microwave vs conventional heating in hydrogen production via catalytic dry reforming of methane. *Resources Chemicals and Materials*. https://doi.org/10.1016/j.recm.2022.08.003

Shi, H., Su, C., Ran, R., Cao, J., & Shao, Z. (2020). Electrolyte materials for intermediate-temperature solid oxide fuel cells. *Progress in Natural Science: Materials International*. https://doi.org/10.1016/J.PNSC.2020.09.003

Singh, R., Oberoi, A., & Singh, T. (2023). Effect of cylindrical heat pipes on electrochemical hydrogen adsorption within a carbon electrode employed in a modified reversible PEM fuel cell. *International Journal of Hydrogen Energy*. https://doi.org/10.1016/j.ijhydene.2023.09.045

Zhou, S., Rasha, L., Xu, L., Du, W., Shearing, P., Coppens, M., Brett, D., & Jervis, R. (2023). Nonuniform compensation of current density distribution in polymer electrolyte fuel cells by local heating. *Energy Conversion and Management*. https://doi.org/10.1016/j.enconman.2023.117717

3 Pre-Heating the Reactant Gases

3.1 INTRODUCTION

In hydrogen fuel cell technologies, particularly in Proton Exchange Membrane Fuel Cells (PEMFCs), the efficiency of electrochemical reactions greatly depends on the temperature and humidity of the incoming reactant gases, specifically hydrogen and oxygen (or ambient air) (Melnik et al., 2024). Pre-heating these gases means raising their temperature in a controlled way before they enter the fuel cell stack. This ensures they are warm enough for efficient operation without getting too hot. This process is especially important during cold starts. Low ambient temperatures slow down reaction rates, reduce the proton conductivity of the membrane, and increase the risk of ice forming in the flow channels or the membrane electrode assembly (MEA) (Zhan et al., 2018).

The primary role of pre-heating is to assist the fuel cell in working faster and with fewer mistakes. It makes sure that the reactants are taken to temperatures that enhance ion transport and electrochemical activity. The optimized fuel cell can deliver a faster start-up, less energy to warm up, and the fuel cell will produce a stable and consistent power output. Moreover, pre-heating prevents inner condensation and accumulation of water. These issues may clog channels through which the gas flows, flood a system, and break parts of the fuel cells. The reduction of the thermal variations in the system with the use of pre-heating can also yield the outcome of a more even distribution of temperature and a diminished number of repeated heating and cooling processes, which will result in less mechanical strain (Moghaddam et al., 2024). In automotive uses and other mobile power systems, quick start and reliability are essential. Therefore, pre-heating the reactant gases is one of the key design requirements. The input energy required to pre-heat should be carefully controlled to avoid reducing the general system performance. Thus, given that the process cannot be carried out at high temperatures and that little energy is available, the process needs to be optimized to raise temperatures in a short time with minimum energy. The following sub-section provides the foundational concepts, relevance, and purposes when considering pre-heating reactant gases (Amamou et al., 2016; Nalluraya et al., 2011). It introduces the illustrative analysis of thermodynamics, technologies, and approaches to the integration of systems that are discussed in the following sections.

DOI: 10.1201/9781003657866-3

3.2 THERMODYNAMICS OF GAS PRE-HEATING

The pre-heating thermodynamics aspects of gases play a pivotal role in comprehending the importance of heating the reactant gases, such as hydrogen, oxygen, or air, entering the gaseous hydrogen fuel cell stack. This is a straightforward transfer of heat by basic heat transfer and energy balance, and how we can deposit heat energy on the gases. This is to increase their temperature either to ambient or above-ambient temperatures, down to the correct operating conditions. Good pre-heating ensures that the reactants are heated to a temperature that allows efficient and stable electrochemical exchange, and this becomes particularly important in PEMFCs, where the temperature strongly influences the performance (Melnik et al., 2024). Without the available thermal energy, the system can malfunction during a cold start or upon transient conditions where the system is started with low temperatures. These are slow response rates, drying of the membranes, elevated internal resistance, and even the formation of ice in the cell or flow pathways. Thermodynamic analysis will allow engineers to determine the precise heat required. They consider parameters such as mass flow rate, specific heat capacity, and temperature difference to attain the desired pre-heating, without wasting a lot of power. Besides, the process should adhere to the second law of thermodynamics. This law shows that there will be a general advantage in minimizing the generation of entropy to enhance system efficiency. The task is to heat in a controlled, efficient, and responsive way. The energy to be supplied must meet the performance requirements of the fuel cell. As such, pre-heating systems require knowledge about thermodynamic principles in terms of design. Such systems must facilitate fast and trustworthy fuel cell start-up and must also have high efficiencies and durability even at various environmental and operational circumstances (Dieterich et al., 2017).

The chemical reaction that occurs between hydrogen and oxygen in a hydrogen fuel cell depends on the temperature. This implies that the rate and effectiveness of the reaction are highly dependent on the thermodynamics of the system. The reaction is slowed at low temperatures, particularly when cold starting. Such a delay can influence the quantity of the electrical power generated. Simultaneously, there is a severe decrease in the ionic conductivity of the polymer electrolyte membrane (PEM) due to reduced hydration and restricted thermal excitation. It can be difficult to control water within the fuel cell as well. Water can solidify or freeze, leading to floods or ice jams. These issues block the flow of gas and may cause the deterioration of the internal structure. To overcome such problems, the feeding reactant gases (hydrogen and oxidants, typically air) must be pre-heated. This raises their temperatures to the optimum operating temperatures of approximately 60–80°C in PEM fuel cells. This temperature treatment increases the speed of reactions, membrane hydration, and water manipulation. The first law of thermodynamics is used to calculate the amount of energy required to heat the gas and is indicated in the following equation:

$$Q = m.c_p.\Delta T$$

where

Q = Supplied Heat Energy (J)
m = Mass flow rate of the gas (kg/s)
c_p = Specific heat capacity at constant pressure (J/(kg.K))
ΔT = Change in temperature (K)

Hydrogen, interestingly, has quite a favorable specific heat capacity, approximately 14.3 kJ/kg K at 25°C. This implies that it takes much energy to increase its temperature per unit weight; this is significantly higher than it does for air, which takes approximately 1.0 kJ/kg. It is due to this that the thermal design of hydrogen pre-heating systems is especially important. The work of this heat exchanger, electric heaters, catalytic heating elements, etc., involves rapid and even heat transfers with minimum energy losses. It is thus required to optimize the thermo-mass of the system, the flow pattern, and the heat transfer area to achieve energy-efficient systems and optimum pre-heating capacity (Sinha et al., 2023).

Pre-heating gas is irreversible by its nature and results in an entropy of some amount, the measure of energy dispersal, or system disorder. Through this entropy increase, the quality of energy or the amount that can be used is lost. It is therefore important to reduce it to improve the overall thermodynamic efficiency of the fuel cell system. The second law of thermodynamics suggests that this principle implies that there is a specific amount of energy needed to warm the gases of reactants. It should be pursued in a way that is as efficient as possible, which also produces the least possible increase in entropy. To fulfill this aim, systems could be designed to be like the reversible transformations, which are highly economical processes that have minimum thermodynamic losses. This can be achieved through the high-efficiency heat exchangers, which maximize the thermal energy transferred between fluids with a minimum temperature difference. Besides, waste heat recovery of elements like fuel cell stacks, coolant loops, or exhaust gases is an energy-efficient and sustainable technology (Egeland-Eriksen et al., 2021). The energy efficiency of the system optimizes with the increase in this recovery, which is the conversion of energy input to floor and thermal energy into useful work (Palma et al., 2013).

3.3 TECHNIQUES FOR PRE-HEATING REACTANT GASES

Hydrogen fuel cell systems need pre-heating of reactant gases, particularly in cold conditions, to provide system reliability, efficiency, as well as longevity. Fuel cells are severely affected during start-up, especially in low temperatures and in sub-zero climates. Such problems include the formation of ice in flow channels, dehydration of membranes, a slow rate of reaction, and the existence of higher internal resistance. These problems may severely impede the initial start-up as well as the stable functioning of the system. To overcome these issues, pre-heating of the incoming hydrogen and oxidant gases to optimum temperatures before they enter the fuel cell stack is worthwhile. Several pre-heating procedures have been developed depending on the requirements of the hydrogen-based fuel cells. Such approaches take into account major performance requirements, including energy efficiency to make sure that little energy is consumed to provide heating; response time to indicate how soon the temperature to be achieved could be reached; integration flexibility that is related to the way in which the pre-heating unit fits into the existing fuel cell designs; compactness of the systems ensuring that the system is suitable to integrate into mobile and space-confined units like cars or portable power units (Lv et al., 2024). Choosing the right pre-heating technique is not a simple decision. It relies on several factors, including the intended application – for instance, whether the system is used in a vehicle, a stationary power plant, or an aerospace system; the temperature requirements, which differ based on the type of fuel cell and its membrane features; and the available power budget, especially important during cold start when the main stack may not yet produce power. In battery-supported systems, electric pre-heating might work well, while in off-grid or remote setups, catalytic or heat recovery methods might be more suitable because they do not depend on electrical energy (Paramanik & Das, 2025).

3.3.1 Resistive (Joule) Heating

Resistive heating, often referred to as Joule heating, is one of the simplest and common techniques to pre-heat reactant gases in hydrogen fuel cell systems. The principle of operation is simple: when an electric current flows through a resisting material, electricity is converted into heat. This heat is subsequently channeled to the hydrogen or oxidant, typically air or oxygen passes it over or around the heating element. These systems commonly use heating materials such as nichrome (a nickel–chrome alloy). Nichrome is highly resistant, thermally stable, and does not oxidize under elevated temperatures. Where there is a need for insulation and the safety and avoidance of contamination of the reactant gases, the use of ceramic-coated metal parts may be found in specific designs (Gao et al., 2024).

Resistive heating has the advantages of rapid response, temperature control sensitivity, and compactness of design. These characteristics make it particularly suited to applications where both space and weight are at a premium, for example, automotive fuel cell systems. It is easy to direct heating based on the electronic sensors and controllers. This configuration enables the system to accurately maintain the desired temperature span. It is very important to keep this accuracy while changing of load and at points like start-up and shutdown. However, it does have a big shortcoming in that the technique needs electricity. This was found to be a problem during cold start when the fuel cell stack is non-functional, with limited energy to be found on board. For resolving this problem, resistive heaters generally use secondary energy sources, such as lithium-ion batteries, particularly in fuel cell electric vehicles (FCEVs). At cold start-up, the battery supplies input energy to power the resistive heaters and bring the system to a sufficient temperature for optimal performance. After the fuel cell stack is warm and generating power, the heating load may be decreased or controlled. From a design point, the resistive heating systems can be installed rather easily with the flow lines or manifolds. They are modularly extendable onto hydrogen and air feed lines. Although some issues related to energy consumption exist, these heating methods are commonly used for pre-heating due to their simplicity, low cost, turn-down capability, instant response, especially in short-term and low-to-medium temperature range cases. Current research focuses on increasing the efficiency of resistive entities, incorporating phase change materials for better heat buffering, and fine-tuning control algorithms to further reduce start-up energy needs.

Luo et al. (2022) performed an experimental study on the cold start performances of fuel cell vehicles (FCVs) with the coolant pre-heating strategy. They analyzed a commercial FCV under the harsh cold environments down to −30°C of air temperature, highlighting performance indicators, such as the output power, heat generation, voltage uniformity, and energy consumption during the cold start. They broke the start-up phase into five phases of operation. The cold start-up operation could be achieved at a temperature as low as −30°C by the coolant pre-heating strategy, but it consumes a lot of time and considerable energy. Results indicated that the endothermic heat generation of the fuel cell stack was essential for the temperature increase after the first pre-heating. This is an indication of the optimization of the stack start-up temperatures to achieve faster transitions. Authors also found that the voltage uniformity decreased at lower temperatures and higher output powers due to temperature, water, and reactant gas non-uniformities – especially near the blind end of the stack. The results indicated that the energy consumed by the PTC heater was mainly consumed in the latter start-up period. They said that switching to self-heating at the right time, better heat transfer efficiency, and waste heat recovery (e.g., pre-heating intake gases) would add to the overall performance of the device. Their work confirms the viability of the coolant pre-heating strategy and offers insights for reducing energy losses, speeding up start-up, and improving system stability in commercial FCVs (Luo et al., 2022).

Figure 3.1 shows the cold start time and temperature rise rate of an FCV at ambient temperatures of −20°C and −30°C over five operational periods (I-V). Overall,

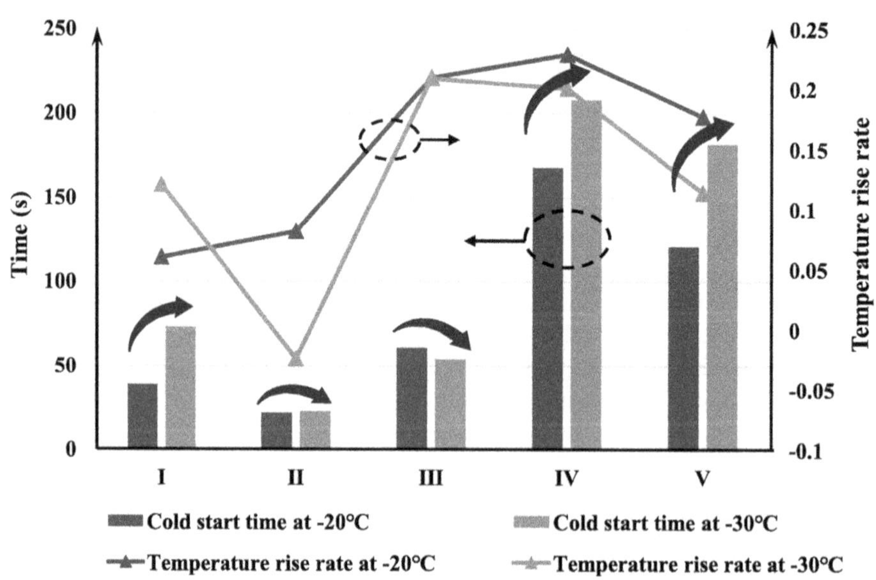

FIGURE 3.1 The time spent in various stages of cold start (Luo et al., 2022).

the cold start time increases a lot at −30°C compared to −20°C, especially during Periods IV and V. This means more effort is needed to raise the stack temperature in extreme cold. The temperature rise rate is usually lower at −30°C, with a significant negative value in Period II. This suggests that heat losses are greater than gains at this point due to colder reactant gases and more heat loss to the environment. During Period I, the coolant pre-heating starts with a PTC heater. Although the rise rate at −30°C seems slightly better at first, it quickly falls when cold hydrogen and air flow into the stack in Period II, absorbing heat and lowering system efficiency. The highest temperature rise rates happen in Periods III and IV when the fuel cell stack starts generating its own heat, but performance at −30°C remains poor due to lower catalyst activity and higher thermal losses. In Period V, the temperature rise rate drops again as the coolant system begins to remove excess heat to keep the operation stable. These changes highlight the important role of self-heating in cold starts. They also suggest that improving insulation, pre-heating reactant gases, and managing thermal conditions at the system level can greatly improve FCV start-up performance in subzero temperatures (Table 3.1).

3.3.2 Heat Exchangers (Indirect Heating)

Heat exchangers are commonly used and energy-efficient devices for pre-heating reactant gases in hydrogen fuel cell systems. They work by transferring thermal energy from a hotter fluid, such as the fuel cell's exhaust gases, coolant loop, or stack surface, to a cooler fluid. In this case, the cooler fluids are the incoming hydrogen and oxidant gases. This transfer happens without the fluids mixing directly. As a result,

TABLE 3.1
Energy Efficiency Outcomes for Different Methods

Method	Cold Start Time Reduction	Energy Efficiency	Temperature Uniformity	Notable Outcomes
PTC + PCM + Self-Heating	19% faster vs. PTC only	Improved	Good	Lower electric heating needed (Zhu et al., 2024)
Coolant Pre-heating	Up to 70% faster (with higher power)	Variable	It can worsen at high power	Successful at −30°C (Luo et al., 2022)
Thermochemical Pre-heater	Rapid, energy-free start	High	Improved	Less ice formation, better voltage (Gießgen & Jahnke, 2023)
Vortex Tube Hydrogen Heating	Raises H2 temperature by ~20°C	Uses tank pressure	N/A	622 W heating power (Lagrandeur et al., 2024)
Sandwich Self-Heating Battery	5.9°C/min heating rate	78.75% efficiency	Good	Used in Beijing Olympics buses (Lin et al., 2023)

heat exchangers enable significant waste heat recovery, which lowers the need for external power sources and boosts the system's thermal efficiency. In hydrogen fuel cells, especially during steady-state operation, a lot of heat is produced during electrochemical reactions. If not used, this heat would simply be wasted through cooling systems or exhaust. By placing heat exchangers strategically within the system, we can capture this thermal energy and use it to pre-condition the reactant gases before they enter the stack. Pre-heated gases speed up reactions, improve water and heat management, and help prevent cold spots that could cause condensation or localized performance issues.

Several heat exchanger designs are used based on space constraints, thermal needs, and system integration requirements:

- Plate Heat Exchangers consist of thin, corrugated plates stacked together to create channels for hot and cold fluid flows. Their compact size and high surface area make them ideal for automotive and mobile fuel cell systems.
- Shell-and-Tube Heat Exchangers are typically larger and consist of a bundle of tubes that carry one fluid, enclosed within a cylindrical shell that carries the other fluid. These are sturdy and suitable for higher flow rates and industrial applications.
- Microchannel Heat Exchangers employ very small flow passages, or microchannels, to achieve high heat transfer rates and quick thermal response. Their small size is especially useful in space-limited or portable fuel cell systems and allows for precise control over thermal profiles.

The effectiveness of these heat exchangers depends on several factors, including the temperature difference between the heat source and the gas, flow rates, material thermal conductivity, and resistance to fouling. In automotive fuel cell systems, for example, heat from the coolant, typically kept around 60–80°C, is transferred to the incoming air stream to raise its temperature close to operating conditions without drawing extra power from the battery or auxiliary heaters. By recycling thermal energy, heat exchangers help reduce the overall energy consumption of the system, which improves fuel efficiency and extends vehicle range. Additionally, the passive nature of heat exchange, with no moving parts in many designs, increases reliability and lowers maintenance needs.

Lagrandeur et al. conducted a new study to improve the cold start performance of PEMFC systems in FCEVs by integrating vortex tubes (VTs) for hydrogen pre-heating. The authors suggested using the high storage pressure of hydrogen (700 bar) to power Ranque–Hilsch vortex tubes, which separate the gas into hot and cold streams. They then used heat exchangers to transfer ambient heat into the cold stream before it enters the fuel cell. They developed a real gas thermodynamic model, along with a multi-objective genetic algorithm, to optimize key VT parameters, such as Mach number, cold mass fraction, and radius ratio, as well as the UA values of heat exchangers for three different system configurations. The study showed that all configurations significantly increased hydrogen feed temperature compared to traditional isenthalpic expansion. The temperature rose from −1.6°C to as high as 18.3°C, with heating power ranging from 325 to 630 W. Notably, Configuration 3, which uses three VTs and two heat exchangers, achieved the best performance but required more pressure and made the system more complex. The authors also found that vortex tube effectiveness remains strong under different inlet pressures and temperatures, and that the design can be scaled for various conditions. The study concludes that vortex tubes can effectively use otherwise wasted energy from high-pressure hydrogen storage to improve cold start capabilities in FCEVs. However, experimental validation is needed to confirm the predictions based on the model (Lagrandeur et al., 2024).

Chang et al. conducted a study to explore the cold start performance of a gas-heating-assisted, air-cooled PEMFC stack. Their goal was to tackle the issues that arise during operation at sub-zero temperatures, where conventional air cooling leads to quick heat loss. The authors began with a thermal balance analysis to find out how much heating power was needed under different ambient conditions. They proposed a hybrid cold start method that combined endplate heating and inlet gas heating, tested this method in a climatic chamber using a 95-cell stack, and observed voltage uniformity and temperature distribution across the stack. Their results showed that gas heating alone was not enough to start a successful cold start at temperatures below −15°C. This was due to the "endplate effect," where extreme thermal inertia near the edges of the stack hindered adequate warming. This resulted in ice formation and voltage collapse in nearby cells. However, by combining endplate heating with gas pre-heating, the stack successfully started at −20°C, showing improved temperature consistency and voltage stability. The study also pointed out that increasing gas heating power improved cell voltage consistency and lowered thermal gradients. Although a cold start was achieved at −30°C using only 14% of the stack's rated power, dropping the ambient temperature to −40°C required unreasonably high heating loads. This indicated a balance between practicality and energy efficiency. They observed that effective cold start strategies for air-cooled PEMFCs must manage thermal conditions while keeping power consumption low. In addition, they also called for more research on flow field design, material selection, and MEA optimization for very low-temperature applications (Chang et al., 2024).

Figure 3.2 shows the heat transfer methods involved in the cold start process of a gas-heating-assisted, air-cooled PEMFC stack. At the center is the stack core, where electrochemical reactions happen and produce internal heat, indicated by arrows moving through the stack. Surrounding the stack are endplate assemblies, which absorb heat through conduction from silicone heating strips. The outer layers are thermal insulation materials that limit heat loss to the surroundings. Even with insulation, heat is still lost through natural convection and radiation, illustrated by curved arrows and by chevrons around the insulation. A PTC heater generates heated air, which flows into the cathode side of the stack via forced convection, represented by arrows entering and exiting the stack. This pre-heated gas is crucial for increasing the stack's temperature during start-up. The diagram emphasizes the combined use of gas heating and endplate heating as effective ways to ensure even and adequate heating across the stack. It also points out the challenges from heat loss and thermal gradients, especially at the stack ends, which can affect start-up performance in sub-zero temperatures.

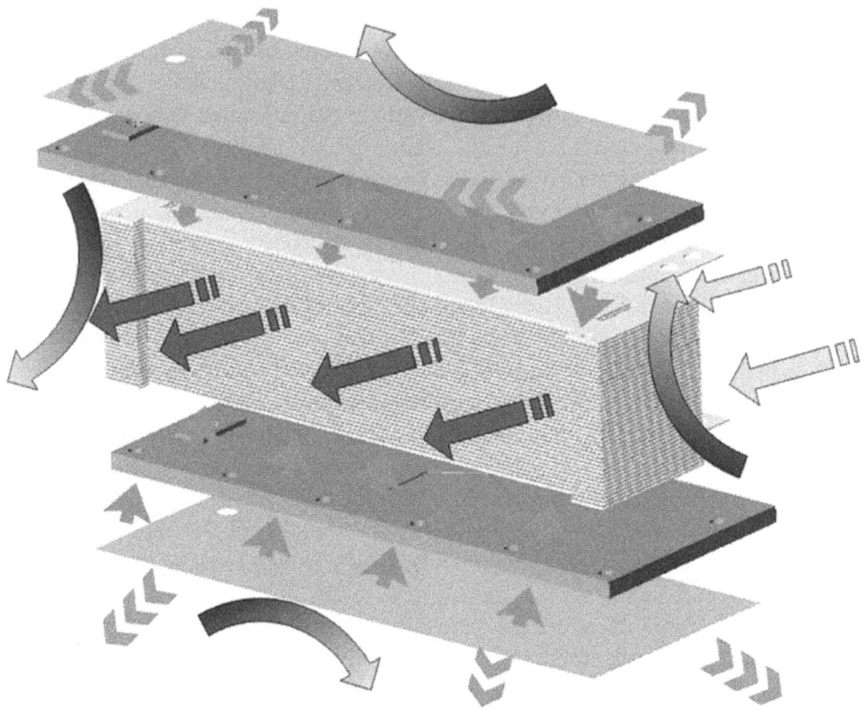

FIGURE 3.2 Heat transfer diagram of the air-cooled PEMFC stack during the cold start process (Chang et al., 2024).

3.3.3 Catalytic Heaters

Catalytic combustion of hydrogen or hydrocarbons generates heat without flames. It relies on a reaction between fuel, such as hydrogen, methane, propane, or butane, and oxygen over a solid catalyst, usually platinum or palladium. Unlike regular combustion, which involves flames and high-temperature areas, catalytic combustion works at lower temperatures, typically between 300°C and 600°C. It occurs without visible flames or loud noises. This process is especially useful in situations where safety, low emissions, and limited space are important.

3.3.3.1 Mechanism of Catalytic Combustion

The process starts when a gas mixture of fuel and air, or oxygen, flows over the catalyst surface. The catalyst helps the heat-producing oxidation reaction by reducing the activation energy needed. This allows combustion to happen on the catalyst's surface. For hydrogen, the main reaction is

$$H_2 + \frac{1}{2}O_2 \rightarrow H_2O + \text{Heat}$$

This reaction is very efficient. It produces only water vapor and a small amount of NO_x, making it an environmentally friendly heat source. The effectiveness and reliability of catalytic pre-heating systems for hydrogen fuel cells depend on careful design and material choices. These systems must provide efficient, controlled, and safe heat transfer to the reactant gases – hydrogen and air – especially when starting in cold conditions. To achieve this, catalytic heaters consist of several integrated parts that work together to improve performance. Catalyst-coated substrates lie at the core of the system. This is where catalytic combustion takes place. Usually, these substrates are composed of ceramic materials such as cordierite, alumina, or metal monoliths, for example, made from stainless steel or Fe-Cr alloy. They provide an increased surface area and temperature stability. The surface is plated with noble metal catalysts, primarily platinum or palladium. They are highly heat-resistant and have strong catalytic activity. This makes it possible to burn hydrogen or hydrocarbons at very low temperatures without requiring a visible flame. The substrate is surrounded by a gas distribution manifold. It also provides a consistent delivery of fuel and oxidant gases over the entire surface of the catalyst. Even heating requires an evenly distributed heat source, as it reduces performance or deactivates the catalyst for all incomplete reactions.

To manage the generated heat effectively, the system uses high-performance thermal insulation materials such as aerogels, ceramic fibers, or composite layers. These materials reduce heat loss to the environment and maintain the internal temperature within the ideal operational range. This improves overall thermal efficiency. For safe and responsive operation, temperature sensors like thermocouples and Resistance Temperature Detectors (RTDs), along with gas sensors for hydrogen, oxygen, and combustion by-products, are placed throughout the system. These sensors send data to a real-time control unit, which adjusts the fuel–air mixture, flow rates, and other operational parameters to maintain the desired thermal output. This active monitoring ensures that the system reacts quickly to changes in demand or surrounding conditions, especially during dynamic cold starts. Choosing the right materials for all components is crucial. While platinum and palladium are the current industry standard because of their excellent catalytic properties, their high cost and limited availability create economic challenges. Therefore, researchers are working on alternative materials like nano-structured catalysts, transition metal alloys, and perovskite-based supports. These new materials aim to keep high activity while being more resistant to sintering, poisoning, and heat damage. This will improve system durability and lower costs.

Ivanova et al. proposed and tested a new cold start strategy for PEMFCs operating at –40°C. This strategy uses a built-in catalytic heating unit integrated into the bipolar plates (BPs) of the stack. The key innovation is embedding three platinum-based catalytic hydrogen recombiners directly into the BPs. This allows direct heat transfer to the MEA through the exothermic reaction of hydrogen and oxygen. This design improved heating efficiency to 60%, allowing the PEMFC stack to reach operating temperature (35°C) in just 6 min using only 1.8 vol.% hydrogen in an air flow of 150 mL/min – well below the autoignition limit, ensuring safety. The authors studied heat transfer models, optimized hydrogen concentration, and looked at how air flow rates affected efficiency. Experimental results showed that heating to only 20°C was not enough for a successful cold start due to MEA dehydration and degradation, while heating to 60°C caused too much dehydration and performance loss. The best condition was heating the BPs to 40°C, resulting in a quick (6-minute) and safe start-up without thermal damage while maintaining water balance and stack stability.

The study concludes that integrated catalytic heating provides a compact, efficient, and controllable solution for deep cold starts of PEMFCs. It outperforms traditional external heating methods and minimizes hydrogen use and component degradation (Ivanova et al., 2023).

3.3.3.2 Advantages of Fuel Cell Applications

1. Flameless and Safe Operation: Since there is no open flame, the risk of accidental ignition or flashback is greatly reduced. This makes it ideal for enclosed environments like FCVs or portable backup systems.
2. High Energy Density: Catalytic heaters produce a lot of heat for their size. This makes them compact and efficient. This quality is particularly useful in settings where space and weight are limited, like in the automotive and aerospace industries.
3. Low Emissions: In contrast to traditional burners, catalytic combustion releases negligible amounts of carbon monoxide, nitrogen oxides, and unburned hydrocarbons. This is in line with the clean energy objectives of hydrogen technologies.
4. Rapid Thermal Response: By a swift approach to operating temperatures, catalytic heaters allow for rapid and efficient pre-heating of reactant gases. This is useful during cold start scenarios where the rapid thermal conditioning of hydrogen and air significantly improves fuel cell performance and durability.
5. Robust in Harsh Conditions: Such heaters function properly in both low temperatures and varying pressures, offering flexibility for electric mobile and field-based hydrogen systems.

3.3.4 Regenerative Heating Systems

Regenerative pre-heating systems are a subclass of thermal energy recovery technologies. They enhance energy efficiency alongside thermal stability in processes with continuous or large volumetric flow rates, such as in fuel cell plants, industrial furnaces, and the function of hydrogen reformers. Different from direct and resistive heating methods, which provide heat on demand via electrical means or catalysis, regenerative systems store and reclaim heat. These types of systems use a thermal storage medium made of ceramic materials with high strength and high temperature performance, like alumina, cordierite, or silicon carbide. Also, metals with high thermal mass and good heat retention structures serve these purposes as well. The filler material is often arranged in the form of matrices like structured honeycombs or packed beds to increase their surface area for gas contact and improve the overall filling factor. In operation, hot exhaust gases pass through these structures, cooling down while transferring part of their energy into the structure, charging the regenerator with thermal energy.

The flow of gas is redirected or reversed when the storage medium reaches a high enough temperature. At this stage, the now-hot medium is passed through cold reactant gases, like air and hydrogen. Before entering the reactor or fuel cell stack, these gases are pre-heated as they travel through, absorbing the heat that has been retained.

The heating of the storage medium and the removal of heat from it alternate continuously throughout this process. For hydrogen fuel cell applications, regenerative pre-heating systems provide a reliable and effective way to preserve thermal stability. These systems pre-heat incoming gases to maintain a steady operating temperature by absorbing and storing exhaust gas heat in materials like metals or ceramics. By shielding delicate parts from heat shock, such as membranes, catalysts, and seals, this thermal buffering improves overall performance and durability. Regenerative systems usually react more slowly than resistive or catalytic heaters, but because they operate passively, they require less energy to start up, have lower operating costs, and require less maintenance. Because of these benefits, they are especially well-suited for Combined Heat and Power (CHP) systems, industrial hydrogen production, and stationary fuel cell installations. Regenerative systems provide significant energy and emissions reductions, recovering up to 80–90% of waste heat. By integrating with automated control systems, heat transfer and overall efficiency can be further optimized through real-time adaptation to operating conditions.

3.3.5 Microwave and Induction Heating

Microwave and inductive heating are being investigated more and more as next-generation techniques for rapidly and precisely pre-heating reactant gases in hydrogen fuel cell systems. These techniques, in contrast to conventional approaches that rely on bulk fluid heating or surface conduction, deliver thermal energy directly to targeted zones in the gas distribution channels, catalyst layers, or metallic flow fields without uniformly heating the entire system. By stimulating polar molecules or conductive particles in the material or gas stream with electromagnetic waves, usually in the 2.45 GHz range, microwave heating generates heat internally through dielectric losses. On the other hand, inductive heating creates localized eddy currents in conductive components, such as metallic heat exchangers or stainless-steel flow plates, by using electromagnetic induction. This results in quick and controllable heat generation where it is needed. These methods are especially useful for cold start scenarios, where rapid and focused heating is crucial for activating the system without using too much energy. Moreover, since these techniques do not require extensive external heating equipment, they help create compact system designs and possibly improve thermal efficiency. While still mostly in experimental stages, initial studies and prototypes have shown their effectiveness and scalability for fuel cell applications. There are challenges in improving energy coupling efficiency, achieving uniform heating in complex shapes, and managing possible electromagnetic interference. Nonetheless, the quick and precise nature of microwave and inductive heating makes them strong candidates for future compact, high-efficiency fuel cell systems, especially in automotive and portable settings where space, weight, and energy limitations are important.

Guan et al. carried out a detailed study to design and assess a catalyst-coated tubular reactor with electromagnetic induction heating for hydrogen production through methanol steam reforming (MSR). Their goal was to reach zero-carbon emissions and quick cold start performance suitable for vehicles and distributed hydrogen stations. The authors created a computational fluid dynamics (CFD) model that integrates electromagnetic fields, heat and mass transfer, and chemical reaction kinetics, using a $CuO/ZnO/Al_2O_3$ catalyst coated inside a stainless-steel tube surrounded by induction coils. By changing parameters such as input current, frequency, coil shape, steam-to-methanol ratio (S/M), and inlet temperature, they found that induction heating allows for precise and rapid temperature control. Reactor wall temperatures exceeded 500 K within 50 s at 50 kHz, which significantly cut the cold start time. They achieved nearly 100% methanol conversion and efficient hydrogen production under the best conditions (e.g., 150–200 A and 15–25 kHz), with higher frequencies enhancing thermal efficiency due to stronger skin effects. They also noted that increasing the S/M ratio and adjusting coil-wall distances can reduce CO generation, which is crucial for PEM fuel cell compatibility. The study shows that induction heating is better than traditional methods because it removes cold spots, improves thermal uniformity, and raises reactor efficiency, making it a promising method for clean hydrogen production in mobile and decentralized energy systems (Guan et al., 2024).

Figure 3.3 shows the design and operation of a catalyst-coated tubular reactor used for MSR. It is heated by electromagnetic induction to produce hydrogen efficiently. The reactor is a cylindrical tube coated on the inside with a $CuO/ZnO/Al_2O_3$ catalyst layer, where the reforming reaction occurs. A mix of methanol and steam is injected from the left side into the reactor, flows through the catalytic zone, and exits on the right side as hydrogen-rich product gas. Surrounding the reactor tube are several induction coils, arranged evenly above and below the tube. These coils create alternating magnetic fields. The fields induce eddy currents and hysteresis losses in the stainless-steel tube wall, generating heat that moves inward to the catalyst layer. This setup allows for quick and even heating of the reactor wall, which is crucial for reducing cold start times and keeping optimal reaction temperatures. The schematic also shows key geometric parameters like the tube's outer and inner diameters (d_0, d_1, d_2), the coil-to-wall distance (b), and the catalytic zone length (l). These factors all affect the efficiency of heat transfer and reaction speed.

3.3.6 Hybrid Heating Approaches

To balance the need for quick system start-up with long-term energy efficiency, hybrid pre-heating systems have become a practical and high-performing solution in fuel cell technologies. These systems combine two or more different heating methods, using the strengths of each to improve performance at various stages of operation.

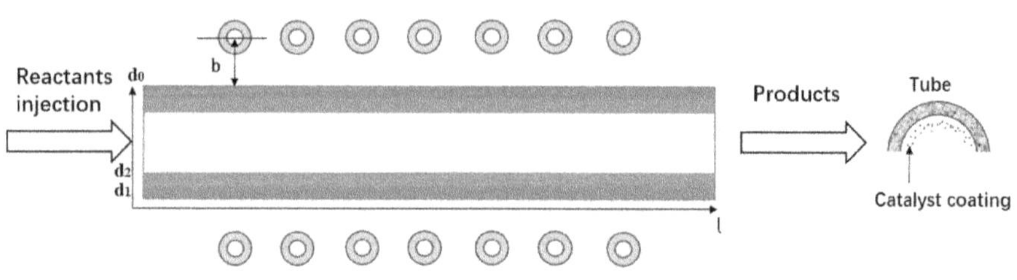

FIGURE 3.3 The reactor physical model (Guan et al., 2024).

For instance, because they produce heat instantly and raise the temperature of reactant gases rapidly, resistive heating elements are frequently utilized during the initial start-up phase. Also, even to be more favored and efficient to avoid having a constant temperature of the gas without, on top of that, the need of actual electrical energy if the system is stable and in a steady state working situation, one can mention in this context heat exchangers and so on that make use of the recovered waste heat of the fuel cell exhaust or any other subsystems. Operating in dual mode also lessens the load of each heating element since they cannot overheat and consume less energy.

These hybrid systems typically use smart control units, which are equipped with automated logic and real-time sensors for controlling heating mode shifts. These systems have the advantage of continuously monitoring environmental conditions, start-up time, gas temperature, and flow rate. This allows the controller to seamlessly blend heating modes or to activate both heating modes simultaneously if desired. Intelligent algorithms, which are often based on machine learning techniques, enhance the responsiveness and adaptability of these control strategies by predicting the ideal heating profiles for different load situations. Therefore, hybrid pre-heating systems provide good thermal efficiency in normal running but are quick and reliable to start from cold. They are best adapted for use in stationary power systems, fuel cells in motion, and hybrid electrics, among other environments.

3.4 DESIGN AND COMPONENTS OF GAS PRE-HEATING SYSTEMS

Hydrogen fuel cell gas pre-heating system design is a challenging engineering undertaking. Thermal performance, energy efficiency, system compactness, and operational reliability must all be balanced. These systems are particularly crucial during cold starts, when the outside temperature falls below the fuel cell stack's optimal operating range, which is typically between 60°C and 80°C for PEM fuel cells. Raising the temperature of incoming hydrogen and oxidant gases before they enter the MEA is the primary goal of a gas pre-heating system. This procedure facilitates stable and effective electrochemical reactions. Heat sources, heat exchangers, flow channels or ducts, temperature sensors, control valves, insulation, and thermal management units are some of the parts of a typical reactant gas pre-heating system. An electrical resistive heater, a catalytic burner, or a heat exchanger that recovers waste heat from other subsystems, such as the exhaust gas stream or fuel cell coolant loop, can all serve as the heat source. Because they are small and react fast, resistive heaters are frequently used in low-temperature applications; however, if not properly controlled, they may result in energy penalties. In contrast, small amounts of hydrogen are used by catalytic heaters to produce heat without burning, offering high thermal efficiency and safety.

To transfer heat to the reactant gases, heat exchangers are essential. Depending on the scale, flow rate, and pressure conditions, designs like shell-and-tube, microchannel, and plate-fin heat exchangers are recommended. The high surface area-to-volume ratios of microchannel heat exchangers make them unique. They take up very little space and enable effective and consistent heat transfer, which is essential for automotive applications. Aerogels or ceramic fibers can be used for thermal insulation to keep gas temperatures constant and reduce heat loss. Temperature and flow rate monitoring in real time requires an efficient sensor and control system. Along the flow path, sensors such as RTDs or thermocouples are positioned to provide data for feedback control loops. To maintain the temperature within the desired range, a PID (Proportional-Integral-Derivative) controller modifies the heater output or flow rate. Adjustments based on start-up conditions, load demand, and ambient temperature are made possible by clever integration with the fuel cell's energy management system (EMS). Components that face high temperatures and reactive gases need to resist corrosion and be thermally stable. Common materials include stainless steels, Inconel alloys, and coated aluminum. Recent research highlights that additive manufacturing (3D printing) of heat exchanger components allows for custom shapes that improve thermal performance and reduce weight, particularly in aerospace and mobile applications. To ensure efficiency and longevity, the design must also focus on minimizing pressure drop, managing thermal stress, incorporating safety mechanisms like over-temperature protection, and creating compact packaging that fits within limited system spaces. CFD and Finite Element Analysis (FEA) are widely used in the design phase to model heat distribution, stress points, and flow characteristics.

3.5 PRE-HEATING IN SPECIFIC APPLICATIONS

Pre-heating reactant gases is designed to meet the specific needs of various hydrogen fuel cell applications. Each fuel cell system, whether used in vehicles, stationary power, aerospace, or portable power, operates under different thermal, environmental, and performance conditions. Pre-heating is not a one-size-fits-all solution; it needs to be tailored based on several factors, including the type of fuel cell technology (like PEMFC or SOFC), the characteristics of the reactant gases, expected ambient operating temperatures, start-up time requirements, and available energy sources. For instance, in automotive applications, quick start-up and compact design are crucial. Therefore, pre-heating systems need to be highly responsive and integrated into the vehicle's thermal management system. In contrast, stationary power systems in remote or cold areas may focus on reliability and energy savings. These systems may require slower but efficient pre-heating, often powered by external grids or auxiliary batteries. Solid oxide fuel cells (SOFCs) also require carefully staged pre-heating to avoid thermal shock to ceramic parts. Onboard reformers used for generating hydrogen from hydrocarbons must pre-heat feedstocks to start high-temperature chemical reactions.

Each application presents its own technical challenges, such as managing thermal gradients, preventing ice formation, reducing energy losses, ensuring even gas distribution, and protecting material integrity. Thus, the design and function of the pre-heating system depend not only on the target operating temperature but also on how quickly that temperature needs to be reached, how it is maintained, and how efficiently the system uses available thermal energy. Often, improved control systems, new materials, and compact heat exchanger designs are necessary to address these varied needs. The following sections explain how pre-heating is implemented and improved in key fuel cell technologies and real-world applications, highlighting the customization needed for different operational situations.

3.5.1 Proton Exchange Membrane Fuel Cells (PEMFCs)

One of the most advanced fuel cells is the PEMFC. They find extensive application in automobiles, handheld power applications, and in the newest homes with residential applications. They are widely used because of their low operating temperatures, fast start-up, and comparatively simple design. They are, however, extremely susceptible to cold weather, which causes some difficulties when making cold starts.

PEMFCs work only in a rather small temperature band of about 60–80°C, normally under special conditions. The proton conductivity of the membrane is constant in this range, and there is stability in the overall water management. At low temperatures and humidity, hydrogen fuel cells experience several performance-limiting issues, including the following:

- Dehydration of membranes is common due to low humidity and low temperatures, at which time ionic conductivity becomes minimal.
- The rates of reactions become much slower, which increases the slowness of start-up and energy consumption.
- The ice may crystallize on the gas diffusion layers or in the flow channels, or even on the MEA.

Reaction times slow considerably, resulting in long start times and high energy use in general in most cases, alas. In certain circumstances, ice may form in the gas diffusion layers or flow channels and sometimes also in the MEA. The processes forming ice can have a terrible effect on inhibiting the conduits of gas exchange and can have a destructive effect on the other fragile components of the cells. Under some conditions, obviously, it is necessary to pre-heat reactant gases, for example, air and hydrogen. *Warming up* the gases before entering the fuel cell stack does a fairly good job of preconditioning the humidifying membrane. It prevents the accumulation of moisture within the stack, hence allowing electrochemical reactions to take place at a brisk rate during normal operating temperatures. Quicker cold starts occur very smoothly and even safely, and lower power bills whilst improving the reliability of the system functioning in general. Electric vehicles with PEMFCs respond crucially at a fast pace. Whether it is the extreme cold outside, drivers still have the urge to power instantly. Pre-heating system warms reactants to the correct temperature in a very short time. In some quite small electric heater schemes, resistive devices incorporated into gas lines or inlet manifolds can warm things up. Heaters should be highly efficient, long-lasting, and compatible with the thermal management system of the vehicle that controls the temperature of the battery and cabin.

3.5.2 Solid Oxide Fuel Cells (SOFCs)

SOFCs are normally used at unbearable temperatures (between 600°C and close to 1000°C). There is a big difference between this and PEMFCs and other comparatively little-known low-temperature fuel cell technologies that operate in quite different conditions. High temperature is essential in the achievement of any conductivity needed by ions in the solid oxide electrolyte, and effectively brings about an electrochemical reaction taking place without any precious metal catalyst. This is the primary reason why rigorous thermal requirements are needed, especially during start-up and at shutdown, quite often in certain scenarios. During a successful start-up of an SOFC system, it is important to pre-heat incoming reactant gases such as hydrogen or methane and combine them with air or oxygen. In SOFCs, pre-heating acts as a thermal shock protector, but in PEMFCs, it primarily boosts the cold start remarkably at diverse operating conditions. When the cold gases are supplied to a hot ceramic fuel cell stack, the differential temperature may damage or crack the ceramic components, especially the vulnerable ceramic components such as the electrolyte, the interconnects, and the seals. Although these materials are pinpoint consistent at extreme temperatures, they are firm and vulnerable to abrupt temperature changes. To avoid this, the pre-heating mode in SOFCs will be planned to be gradual and well-supervised. It contains a multi-step heating process. To begin with, the temperature of the fuel cell encapsulation and the gas delivery tubings can be increased by external resistive heaters or even by radiant ceramic heating. There is the use of recuperative heat exchangers as the system heats up. These are waste heat recovery gadgets that tap into the exhaust gases of the already in operation stack or virgin burner and apply it toward the pre-heating of the incoming fuel and air supplies. In addition to reducing the start-up time, this technique will also increase the overall efficiency of the system, as the waste heat that would have otherwise been wasted is reclaimed instead.

The smooth integration of pre-heating into the thermal loop is configured easily in most SOFC systems, and particularly the ones that are used in CHP to serve industrial and non-industrial homes. These systems are designed to make electricity and valuable heating of the space or water, radiated by heat, created by the fuel cell stack simultaneously. Pre-heating of reactant gases also becomes a component of the heat recovery strategy and enhances the efficiencies with the use of fuel and its costs of operation. More sophisticated SOFC set-ups can even involve a set of automated temperature ramp-up procedures, wherein temperature increase is regulated by the use of sensors and programmable controllers. This makes sure that all components of the system are not subjected to sudden and uneven temperature changes. This kind of control is needed to realize even temperature distribution, to prolong the life of fuel cell parts, and to ensure the reliability of systems.

3.5.3 Onboard Fuel Reforming Systems

Hydrogen is not supplied or stored as a compressed gas in several systems based on hydrogen energy, especially in generating systems that are concerned about being compact, flexible in fuel selection, and power generation on demand. On the contrary, it is generated in situ by using fuel reforming. It happens in all onboard hydrogen generation systems, where hydrocarbon feeds, such as methane, ethanol, methanol, or even diesel, are used as starting materials. The chemical reactions convert such fuels to gaseous streams of hydrogen-rich products, for example, steam reforming, partial oxidation, or autothermal reforming (ATR). These rearrangements consume quite a lot of heat, thereby making them very endothermic. For example, Steam Methane Reforming (SMR) is typically run at a temperature range of 600–850°C, whereas methanol or ethanol reforming could require a temperature range of

200–350°C, depending on process configuration. Irrespective of pre-heat strategy, the gases that comprise the reformer feed in most cases are a mixture of liquid fuel and water vapor or steam, and need to be pre-heated to commence and continue the reaction. When the feed gases are introduced into the reactor at lower temperatures, the reaction will occur very slowly or it will not occur at all. This may lead to reduced hydrogen production, carbon formation (coking), or catalyst poisoning.

Fine pre-heating solutions are inculcated in the reformer system to address these difficulties. The most popular ways of heating include the following.

3.5.3.1 Catalytic Heating

The catalytic heating technique entails combustion of a small fuel mass, in correspondence with a catalyst, to generate the required sequestration. It is usually more effective and cleaner compared to conventional combustion. This is a usual practice in portable and low-emission systems.

3.5.3.2 Resistive (Electric) Heating

In the resistive (electric) heating method, there are electric heating elements that are used to heat the inlet gas stream to a desirable temperature. It has tight temperature control, which is essential in dynamic load-following or start–stop mobile and auxiliary applications.

3.5.3.3 In Higher-Performance Systems, Pre-Heating of the Feed Gases Is Used

The waste heat of the reforming reaction or the exhaust of a fuel cell can heat the feed gases using heat exchangers in such systems. This enhances the entire energy efficiency. In ATR or hybrid reforming and fuel cell, thermal integration is vital to reach self-sustainability operation, and therefore it is particularly important.

Applications of reformer-integrated pre-heating systems include the following:

- Auxiliary Power Units (APUs) in heavy-duty vehicles, aircraft, or ships. Here, diesel or natural gas is reformed on demand to power fuel cells when the engine is off or during peak loads.
- Portable Hydrogen Generators use methanol or ethanol cartridges to supply hydrogen to fuel cells for military or remote-field applications. In these compact systems, pre-heating needs to be efficient, quick, and independent, as external power sources may be limited.
- Marine Fuel Cell Systems rely on onboard reformers to convert diesel or LNG to hydrogen for fuel cell propulsion or auxiliary energy. Due to space and thermal limitations on ships, these systems depend on smart thermal integration and compact pre-heating units to operate reliably.

Accurate thermal control mechanisms are also necessary for the successful use of pre-heating in these reforming systems. This includes thermal insulation to stop heat loss, PID-based feedback loops, and real-time gas composition analysis. Given the corrosive and high-temperature conditions found in reformer environments, choosing the appropriate materials is essential. Pre-heater components are frequently made of ceramics, high-temperature alloys, and corrosion-resistant linings.

3.5.4 Aerospace and Defense Applications

Fuel cell systems are increasingly being used in the aerospace and defense industries for high-altitude energy systems, military ground transportation, remote surveillance stations, unmanned aerial vehicles (UAVs), and portable power units. High performance, fast activation, and robust resistance to challenging conditions are necessary for these applications. Ensuring dependable operation in extremely cold or quickly fluctuating temperatures is a major challenge, as conventional batteries and power systems frequently malfunction or perform poorly in these conditions. Pre-heating systems are essential for fuel cell operation under these circumstances to address these issues. Water vapor in the reactant gases condenses or freezes when ambient temperatures drop significantly below 0°C in cold climates or at high elevations. Long start-up delays, membrane dehydration, and clogged flow channels can result from this. Fuel cells might not function properly when they are most needed if they are not pre-heated.

In these applications, pre-heating systems are made to respond thermally very quickly; they can frequently raise the temperature of the reactant gas within seconds of activation. In critical missions, where delays are unacceptable, this is crucial. The pre-heating components must be small, light, and energy-efficient. Consequently, popular options include thermoelectric heating elements based on the Peltier effect, resistive microheaters, and even catalytic microburners. To provide effective heating with little heat loss, these devices are positioned near the MEA or along the gas inlet lines. Compound thermal insulation covering, such as ceramic fiber blankets and, more rarely, aerogels, is used relatively commonly nowadays to avoid major heat loss. Insulation enhances the preservation of thermal energy by the heated gases, hence providing sufficient energetic supply to the fuel cell stack thereafter. The total power requirements placed on the onboard power source or auxiliary battery are reduced, thus enhancing the system efficiency by quite a margin. Smart control, automation algorithms are quite important in pre-heating systems extensively applied in aerospace and defense industries. These controllers continuously adjust the heater output in response to real-time sensor measurements of the environment, such as humidity and temperature levels at different pressure circumstances. Reactant gases flow continuously, and the energy is saved, and there is a general avoidance of overheating here, rather efficiently. Autopower systems in different UAVs and autonomous vehicles tend to automatically engage when the power of a vehicle soars or the environment gets very inconsistent.

Durability and redundancy are also required where parts of defense systems have high reliability and survivability requirements. Pre-heating installations are constructed with sturdy materials and vibration-resistant mounts to take the tough blows and endure considerable temperature variations. During pre-heating, the fuel cell could be operated under increased stealth and mobility in a very short time without undertaking extensive external warm-up processes or external aids. Such designs are particularly useful in vehicles used by the military in the freezing Arctic or the baking desert at the same time. The atmospheric temperatures are very low, dropping to −50°C or even lower in such high-altitude locations as balloon platforms or very advanced reconnaissance drones. Pre-heating systems avoid the freezing of internal condensation to create an operational hitch and, at the same time, pre-sublimates the incoming gases to prepare them within the heat. Fuel cell stacks allow longer flight times and more energy-efficient missions by means of clever waste heat use designs.

3.5.5 Stationery and Backup Power Systems

Fixed fuel cell systems must perform completely dependably in very extreme and unfavorable environmental conditions typical in radio communication towers and remote industrial facilities. Systems are usually installed in places that are poorly located and inaccessible, with little probability of human repair or periodic maintenance. Even in Arctic regions or locales that are very cold with sub-zero temperatures, fuel cells need to restart themselves in quick succession after extended dormancy. Maintaining adequate fuel cell system performance with cold start-up is generally a significant challenge. When temperatures are below freezing, flow channels and valves, and sensors may condense water or become covered in ice unless adequate winterization has been done. It can lead to membrane dehydration or mechanical blockage and, in some cases, cause problems with electrolyte balance. Such situations may even break fragile system parts and considerably reduce operation time due to the drastic delays of the start-up process. Operational solutions to these problems require reactant gases such as hydrogen and air or oxygen to be pre-heated. Gases being introduced into the stack should be heated adequately. This prevents challenges that are caused by low temperatures, and it establishes optimal reaction conditions in a short time in diverse operating conditions. Most stationary systems now use electric resistive heaters. Such heaters have high rates of thermal response, which are quite reliable and work quite straightforwardly in most situations. They are typically located within stack enclosures or along gas inlet lines, somewhere near some other important equipment in some way. These heaters are frequently powered by auxiliary generators or separate batteries in an installation with no stable grid connection, and sometimes by a sun-charged energy storage battery. This arrangement permits pre-heating even when the primary fuel cell is not functioning optimally or is cold due to some conditions. Battery life is conserved and start-up speed is retained due to the careful automation and energy-efficient optimization done in advance through pre-heating that happens more or less non-obtrusively. Modern systems have another interesting aspect of incorporating weather-based predictive controls in a considerable number of areas today. Currently, weather forecasts in real time and ambient temperature sensors are used with seasonal data in fuel cell EMSs in a rather successful way. Pre-heating also happens automatically when it detects a lowering of temperature or perceives that such a change may happen more quickly in some circumstances. Such an arrangement guarantees that a system receives a demand signal: it is heated enough and is now ready to operate appropriately. It is a proactive preventive that can greatly increase system preparedness in case of emergency or under heavy load, and reduces delay at start-up. When the fuel cell start-up fails, expensive downtime arises in a telecom infrastructure that must always remain powered on to maintain network availability. More importantly, in outlying installations or wildlife monitoring sites, human intervention can be delayed days or several weeks, and start-up cold.

This is ensured with the pre-heating that keeps the stress out when the temperatures are falling and skyrocketing in this mad, constant swing of temperature, so extreme changes in season can also be relied upon. The option of smart diagnostics that can be combined with a pre-heating system allows real-time monitoring of heater activity and temperature profiles as well as the overall health of the system, which works very efficiently. Predictive maintenance is assured, and this approach substantially reduces the possibility of failure at critical stages of operation.

3.5.6 Fuel Cell Buses and Heavy-Duty Vehicles

Gas-powered buses, deliveries, and trucks are about to die, as fuel cell journeys are suddenly catching up with nationwide diesel fleets in the passenger transportation and cargo delivery sectors. The promise of these vehicles is their outstanding reliability and quick response times in both very cold winters and really hot deserts, with a rather long range. Fuel cell systems technically have a difficult challenge of cold starting and temperature management in cold weather or during start-up in the early morning hours. Pre-heating the reactant gases (hydrogen and air/oxygen) is essential for several reasons. At low temperatures, the electrochemical reactions in the fuel cell stack slow down significantly. Water may condense or freeze in gas channels, and membrane hydration can decrease, which leads to performance loss, mechanical stress, or even stack failure. Pre-heating helps ensure that the reactants enter the stack at a temperature that supports optimal reaction speed and membrane conductivity. This allows for smooth, reliable, and quick system start-ups. However, in vehicles like buses and trucks, the pre-heating system must be effective and energy-efficient. Unlike stationary systems, mobile vehicles have limited auxiliary power available. Excessive energy use by pre-heaters can drain the battery or reduce the vehicle's driving range. Therefore, the pre-heating strategy needs to carefully balance speed and efficiency.

To meet this need, modern FCVs use a combination of three main pre-heating methods:

1. Electric (resistive) heaters are installed in the gas supply lines or inlet manifolds. They provide rapid and precise heating, especially during the initial start-up. The vehicle's EMS controls them to ensure minimal power use and short activation cycles.
2. Catalytic pre-heaters offer a more thermally efficient solution. They combust a small portion of the hydrogen fuel catalytically (without flame) to generate heat for warming the incoming gases. These systems are compact, safe, and produce very little NOx, making them suitable for automotive use.
3. Heat exchangers recover waste heat from the fuel cell stack or exhaust and transfer it to the incoming reactant gases. This method is highly efficient and is essential during continuous operation because it helps keep optimal stack temperatures with minimal additional energy input.

All these pre-heating components are closely integrated into the vehicle's thermal management system, which also manages cabin heating, cooling, and waste heat routing. Intelligent control systems oversee the sequencing and coordination of these pre-heating devices based on ambient temperature, start-up status, and load demand.

In urban transit systems, where buses start early and operate for long hours, reliable pre-heating systems ensure cold start performance meets regulatory and operational standards. In logistics, where trucks may idle for long periods or operate in different climates, efficient pre-heating helps cut cold start emissions, protect the fuel cell stack from thermal stress, and ensure timely deliveries. Pre-heating also contributes to increased passenger safety and comfort. An efficient pre-heated system start-up rapidly warms the passenger area, enhancing the cold-weather experience as the fuel cell also heats the cabin quietly. Manufacturers and researchers are investigating novel techniques like predictive pre-conditioning algorithms and smart heat storage alongside phase change materials thoroughly nowadays. These advancements aim rather quietly to boost the reliability of fuel cell tech and spur its adoption rapidly in commercial transport sectors.

3.6 ENERGY EFFICIENCY AND OPTIMIZATION

Hydrogen fuel cell gas pre-heating systems need to be designed with energy efficiency optimization in mind, especially under extremely cold start conditions. Mishandling pre-heating reactant gases can drastically reduce fuel cell system efficiency by wasting considerable energy unnecessarily during operation. Methods for curbing energy expenditure are examined here alongside strategies for maintaining efficacious thermal conditioning of gas.

3.6.1 Utilization of Waste Heat

Waste heat reactant gas pre-heating is one of the well-established techniques for energy efficiency improvement of hydrogen fuel cell systems, especially during cold start operation. Entering hydrogen and air are pre-heated by the waste heat of the fuel cell stack, the coolant loop, or the exhausting gases. Super rare microchannel-type or fancy plate-fin-like heat exchangers transfer heat between torrid streams, for example, of coolant or exhaust, into cold reactant gas streams without direct mixing of the two. Regenerative heating loops that are used in some designs contain a process fluid that has good conductivity, hence it is routed within processing systems to effectively transfer thermal energy. Temperature sensors and valves that pre-heat and prevent the overheating of the same fluids are quite common and work rather well in systems. Waste heat replaces electric heaters or auxiliary burners, thus greatly streamlining system start-up at a lesser initial energy demand. This strategy is efficient in reducing fuel consumption and emissions and developing sustainability objectives relatively quickly. In automotive PEMFCs, it is quite possible to use a coolant up-regulated to 60°C or 80°C to somewhat effectively pre-heat through otherwise incoming reactants. This technique, more effectively than any other, saves energy reserves in the onboard spacecraft systems; in fact, it reduces cold start times extremely quickly. When incorporated within waste heat recovery, thermal management can be increased substantially and allow the design of smaller systems overall, extending the lifetimes of system components.

3.6.2 Adaptive Control Algorithms

Dynamic control of heater output in the gas pre-heating systems of the hydrogen fuels offers significant energy efficiency increases under the majority of working conditions. Intelligent control strategies applied in systems adjust heating depending on the operating condition in real time instead of predetermined settings. PID controllers constantly gnash temperature and adjust heater output, preserving sweet spots in electrochemical reactions, particularly cold starts. In the case of more involved scenarios, Model Predictive Control, or MPC, comes into the picture quite elaborately. MPC applies arcane mathematical formulae actively in forecasting future behavior of systems and thus balancing wildly varying energy consumption and thermal needs. AI is perfectly integrated into the advanced applications that study work patterns and improve control in incredibly low-level ways, reducing energy waste. Such smart control methods have major advantages. The gases remain within desirable temperature levels because they significantly limit both overheating and underheating, which makes them quickly respond to unstable loads and environment variations and minimize energy losses immensely. In FCVs, an intelligent system identifies self-sustaining temperatures early and lowers the pre-heater load, saving battery energy yet keeping the system operating effectively as a whole.

3.6.3 Advanced Heat Exchanger Design

Under cold start conditions, a well-designed heat exchanger can significantly enhance the energy efficiency of gas pre-heating schemes in hydrogen fuel cells. Transfer as much heat as possible to the reactant gases in a short time with minimum energy losses and a sufficiently modest overall footprint. More modern applications incorporate some fairly esoteric types of microchannels and plate-fin and printed circuit heat exchangers commonly known as PCHEs. The surface area to volume ratio is increased by several orders of magnitude in microchannel exchangers with closely spaced small-diameter parallel channels within very compact configurations. This makes it possible to have a quick and even heating and ensures a minimal level of thermal resistance. Plates and fins are arranged in a crossflow or counterflow arrangement, which increases the rate of heat transfer significantly further. The diffusion bonding produced PCHEs are known to be strong and provide effective heat transfer even within very demanding operating conditions. Baffles or vortex generators are added in a very random fashion today to add some turbulence and vastly increase the rate of thermal exchange. Such improvements assist in the reduction of pre-heating durations to a minimum and conditioning of gases before they get into fuel cells in a remarkably short time today. Integrated solutions and touch-responsive systems are especially beneficial in highly space-constrained environments, such as automobiles, which is a common example. In some applications, clearly sophisticated CFD modeling is essential. The CFD tools will precisely tune exchanger designs to minimize pressure drop without compromising the thermal effectiveness that can reach up to 90%. A microchannel exchanger that was effectively embedded in a PEM fuel cell system was able to attain fast heating using waste heat and cut down cold start dramatically by 30% whilst retaining astonishingly high energy efficiency.

3.6.4 Improved Thermal Insulation

A major goal of enhancing energy efficiency in hydrogen fuel cells is to achieve adequate thermal insulation during cold start-up and low-temperature conditions of fuels. This is the major task, as the whole idea consists of preventing heat loss through scalding gas lines and pre-heating components in the outdoors, and conserving energy. Several high-technology insulating materials are used, accordingly increasing overall performance with properties that can be somewhat exotic. The use of aerogels, which boast the lowest thermal conductivity ever recorded and a ridiculously low density, can be effective in insulating hydrogen supply lines to reduce heat loss by conduction. Vacuum Insulation Panels (VIPs) have a very high insulation capability in physically constrained applications due, in large part, to an evacuated core sealed gas-tight. Heat exchangers and heaters are commonly highly efficiently insulated with ceramic fiber blankets due to their uncooperative nature with respect to high temperatures. Housings and insulated covers surrounded by heaters and heat exchangers reduce radiative losses of heat to a great extent and reduce external cooling quite effectively. Other systems also add multilayer insulation/super-insulated enclosures that act as a protective barrier and a good thermal insulator at the same time. Together, these strategies present a number of vital advantages, including the ability to keep gas temperatures over greater lengths and a significant reduction of the overall energy requirements. They reduce the thermodynamic stresses and variability of performance through the accumulation of heat internally in the system, thus potentially increasing the duration of operation of fuel cell parts significantly. This sort of methodology is clearly seen in the electric vehicle hydrogen fuel cell version. Aerogel-insulated hydrogen lines preserve temperatures of pre-heated gas of about 50–60°C as gas passes through the pre-heater to the stack inlet. Without this insulation, significant heat loss may be experienced, thus increasing the energy responsibility and temporarily halting the start-up of systems. Good insulation will reduce temperature losses by a tremendous amount, as only a 2–3°C decrease over many meters of tubing with cold starts is significantly better.

3.6.5 Pre-Heating during Critical Phases Only

Pre-heating in crucial stages is just an indication of a customized approach to enhance the effectiveness of hydrogen fuel cell systems, by limiting the use of pre-heaters to only necessary times. The system utilizes the pre-heater primarily to perform a start-up, and only briefly in rare operational periods where the thermal environment is not properly conducive to an optimal electrochemical performance. Cold start conditions of sub freezing ambient temperature degrade stack performance because of cold gas reactant streams and the potential formation of condensation or ice. Self-installed pre-heating mechanism comes into play automatically in the first start-up protocol, and the sturdy problems are resolved successfully. The vigorous heating of incoming hydrogen and oxidant gases is adequate to sustain membrane hydration and enable intended electrochemical reactions successfully. Gas inlet and outlet temperature sensors communicate live data to the control unit when the internal stack temperature reaches the operational point of around 65–70°C. Control systems then identify precise intervals at which they can turn off pre-heating capabilities, thus reducing unnecessary energy consumption within the shortest time possible in the majority of situations. This point onward, heat produced by the reactions in the stack and the coolant loop is enough to sustain the designated thermal conditions. The use of this selective pre-heating method has a lot of significant benefits that can be very efficiently achieved nowadays in a variety of industrial applications. Savings of energy are made primarily through the prevention of continual use of heaters, especially useful in the case of FCVs, as well as other portable applications where energy reserves are naturally inadequate. Minimizing temperature cycles and stress to heater elements can significantly prolong service life, as well as make an impact on the system durability that can be considered quite remarkable. Nothing is pre-heated, unless the tube is not heated in the first step of ignition; hydrogen-powered off-road buses are generally used in cold temperatures. When the stack thermally drifts have subsided, the system survives solely on endogenous waste energy to sustain thermic activities. Such methodology increases energy usage efficiency to the maximum extent possible and enhances fuel cell system durability somewhat mysteriously, in addition to component lifetime. Selective pre-heating is therefore a key ingredient that can drive effective cold start capabilities of the modern hydrogen fuel cell technology today.

3.6.6 Computational Modeling and Simulation for Optimization

CFD has evolved into conventional use in the design of gas pre-heating systems for the hydrogen fuel cells. It allows engineers to study the flow and thermal conditions of hydrogen and air through pre-heating channels and heat exchangers without constructing costly prototypes. With CFD, school building problems such as thermal hot spots, inequalities in flow distribution, and pressure drops can be identified at an early stage of the design, which enables most corrections that save time and cost. This is augmented by the FEA that assesses structural issues like thermal stress, deformation, and fatigue at multiple operating conditions. Other start-up situations, such as cold start, are also simulated to optimize pre-heating approaches to bring about rapid, efficient temperature increase without damaging parts or wasting energy. Optimization methods, such as Design of Experiments, are integrated in creating a better refiner of the aspects, such as fin geometry and placement of the heater, that can improve the heat transfer with little energy consumption. Design accuracy is highly increased by this procedure of simulation-based modus operandi, and fewer physical models are required, and the timeline of development is much quicker. It also offers forecast access to the system performance in severe conditions, increasing reliability and performance. A recent study of CFD-based design of plate-fin heat exchanger in PEM fuel cells improved the heat transfer by 20%, power dissipation by 30%, with enhanced thermal uniformity and lower forces on the heat exchanger.

3.7 MATERIALS AND CONSTRUCTION

Water fuel cell, the mode of working is a complex issue in designing a gas pre-heating system when exposed to various conditions, which involves durability, thermal efficiency, and dealing with safety demands at the same time. The materials should have good thermal conductivity and must be highly mechanically stable in the reacting conditions of operating environments involved with fuel cell operating environments, dynamically. Inconel and aluminum alloys are frequently picked for a favorable balance of strength, heat tolerance, corrosion resistance, and stainless steel stays widely used. Ceramic composites and graphite-based materials offer pretty viable alternatives in applications prioritizing minimization of weight and maximization of compactness heavily. Surfaces exposed to hydrogen and oxygen are often treated rather liberally with anti-corrosive coatings, mitigating degradation and preventing hydrogen embrittlement somewhat effectively. Insulation materials like aerogels and ceramic fibers are frequently employed in systems to minimize thermal losses and enhance energy efficiency greatly. Fabrication of super-sophisticated components like microchannel heat exchangers and integrated heater modules demands extremely high precision manufacturing processes nowadays. Techniques like CNC machining and laser welding, alongside increasingly prevalent additive manufacturing, enable the production of complex geometries and compact assemblies tailored for hydrogen fuel cell systems' unique requirements.

3.7.1 Stainless Steel

SS316 grade stainless steel is currently used in building pre-heating gas systems in hydrogen fuel cells with a reasonable degree of success. SS316 is extremely mechanically strong and very thermally stable, perhaps because of the simplicity of its internal composition and its noncorrosive nature. It is used for heater casings, heat exchanger tubes, flow channels, and gas connectors, wherever it encounters a reactive environment or temperatures over 1800 degrees. SS316 performs incredibly well even under high humidity conditions, where water vapor can react rather actively with the acidic byproducts of the fuel cell reactions. SS316 is still very resistant and will have an extremely long life in such environments, thus ensuring excellent long-term safe operation for extended periods of time. This is largely because it is tough under diverse operating conditions and contains molybdenum. Adding molybdenum also improves the chloride resistance of this environment, with remaining electrolytes or small contaminants more frequently. SS316 retains its integrity exceptionally well at temperatures near 800°C and is thus superbly well-suited to some heat exchangers. As part of a focus on efficiency, exhaust streams or coolant loops on critical equipment are surrounded by reactant gases to capture this heat. This cycle of thermal stressing ensures that the system has a long life, and conducts heat reliably without major deformation setting in.

3.7.2 Inconel

Inconel 625 surfaces as a very significant material in other limiting, beyond-boiling applications, such as in a corrosive environment with high design-stress corrosion concentrations or chemical resistance in terms of hydrogen oxidation and embrittlement. It is quite valuable in elements that experience severe thermal cycling conditions in hydrogen fuel cell technology in gas pre-heating systems. This critical property is observed in cold start operations where reactive gases are swirled about, creating major temperature variations within a short period. Inconel 625 is a tenacious material that stubbornly retains structural integrity and significant mechanical strength, hence promising unusual durability and reliable system performance. A major application of Inconel 625 in such industrial processes includes the fabrication of combustion chamber liners and housing of catalytic burners. Critical elements aid in pre-heating hydrogen at ideal temperatures without impurities that might drastically reduce the efficiency of the fuel cell stack. The excellent high-temperature combustion stability and a non-reactive surface of the alloy combine to make it highly suitable in catalytic systems where a stable source of uncontaminated heat is vital in the successful functioning of fuel cells without undue risks to people. Inconel 625 is highly resistant to chemical degradation caused by water vapor and other byproducts, making it ideal under tough conditions in mobile and portable fuel cells as well as stationary fuel cells.

3.7.3 Aluminum and Aluminum Alloys

Aluminum is a key material for developing lightweight compact heat exchangers and is often used in automotive HFCEs where compact assembly is necessary. Its high thermal conductivity also allows for rapid, uniform transfer of heat into incoming reactant gases. Rapid thermal response reduces start-up delay and enhances system readiness and fuel cell performance advantageously as a whole. The excellent strength-to-weight ratio of aluminum has a very beneficial effect in many vehicle applications, and, in general, the advantage of light weight leads to better fuel economy. In some vehicles, like the Toyota Mirai, aluminum-finned microchannel heat exchangers are used to pre-heat incoming air that enters the fuel cell stack. This construction provides for rapidly achieving the desired air temperature in the device and efficient electrochemical reactions to occur with very little additional weight. Aluminum parts can obtain the necessary corrosion resistance by surface treatments such as anodic oxidation or using a protective coating in the presence of moisture. Structural integrity is well kept, and the life of critical parts is, generally, extended quite significantly with time. Somehow, aluminum still wins the day when it comes to advanced pre-heating systems on fuel cell systems on mobile rigs.

3.7.4 Ceramic Composites

Silicon carbide and alumina ceramic composites play a crucial role in the fabrication of high-temperature components used in pre-heating gas systems for hydrogen fuel cells. Rare materials are typically picked characteristically for super-hot insulator jobs and electric heating elements inside non-metallic heat exchangers under extreme thermal stress. Remarkable thermal stability characterizes these ceramics, enabling reliable operation at scorching temperatures exceeding 800°C in advanced fuel cell systems such as SOFCs. Ceramic composites remain remarkably inert chemically. Remaining largely unscathed by hydrogen and oxygen gases, this property averts corrosion embrittlement and untoward chemical reactions that spoil system efficiency or foul fuel cell surroundings badly. Silicon carbide heating elements are often utilized in electric pre-heaters that condition air streams entering

SOFC stacks effectively. These elements withstand significant thermal loads quite effectively and demonstrate impressive durability across numerous repeated heating cycles with rapid thermal response. Silicon carbide proves optimal for high-purity applications due to its remarkable chemical stability and longevity, thereby supporting consistent thermal performance and essential chemical integrity in fuel cell systems for efficient electrochemical conversion.

3.7.5 Graphite and Carbon–Carbon Composites

Graphite and carbon–carbon composites hold a vital area in contemporary gas pre-heating methods primarily owing to exceptionally high thermal conductivity. Attributes like these make them super valuable for creating customized heat sinks and gaskets in applications with serious spatial constraints. Their capability to conduct electricity permits them to be pretty suitable for systems needing incorporated heating elements or mitigating electromagnetic interference pretty efficiently. Graphite exhibits extraordinary resilience in bad chemical environments (as well as other similarly aggressive environments) encountered by hydrogen fuel cells. Material longevity and system dependability are ensured by its persistence in withstanding degradation within extremely reactive environments remarkably. The material's superb thermal conductivity facilitates fast heat distribution pretty uniformly, and this stops localized cold spots or uneven heating. Compact heating modules in transportable fuel cell systems and automobile applications utilize graphite foam or carbon composite structures as heat spreaders. Components crucial for effective pre-heating of reactant gases enable faster start-ups and consistent performance under diverse thermal operating conditions. Comparatively low density of carbon-based materials aids in lowering total system mass, a crucial consideration for different mobile applications, obviously. Graphite and carbon–carbon composites provide substantial versatility in design rather usually with ease. The parts may be machined or additively produced to complicated geometries to enable incredibly effective space-saving thermal management components for precise fuel cell configurations. Graphite is crucial in thermal management strategies because developments in fuel cell technology drive the need for smaller system designs.

3.7.6 Aerogels

Silica-based aerogels indicate a sizable leap in high-performance insulation materials, particularly within hydrogen fuel cell applications involving gas pre-heating methods. They mainly function by minimizing heat loss from heater casings, and gas pre-heating modules are vital elements during cool starts and protracted standby. Silica aerogels exhibit extremely low thermal conductivity at 0.015 W/m·K, significantly exceeding standard insulation components in functionality these days. Outstanding thermal resistance enables designers to lower the size of pre-heating units without sacrificing containment in compact environments like vehicles. These aerogels stay remarkably buoyant and water-repellent, therefore precluding moisture seepage in steamy gas flows and sustaining insulation efficacy as time passes efficiently. Stationary fuel cell systems frequently utilize silica aerogel blankets enveloping pre-heater devices, hence mitigating energy losses during start-ups at lower temperatures. Warm-up phase accelerates rapidly, and system energy demand decreases considerably under cold conditions, ultimately contributing to extended lifespan and enhanced operational efficiency.

3.7.7 High-Temperature Elastomers

High-temperature elastomers, such as very resilient Viton or wonky Kalrez plastic, play a crucial part in creating fuel cell systems. Such materials are indispensable in sealing components, gaskets, and flexible connectors abound, particularly at interfaces where heating elements encounter inlets of PEMFCs. The main reason behind selecting these elastomers stems mostly from a quite unique blend of thermal stability and remarkable mechanical flexibility. The combination of properties ensures system integrity in standard operation and difficult situations like cold starts under incredibly harsh conditions. Pre-heated gases, notably hydrogen, can produce considerable pressure and high temperature at different joints in addition to countless interfaces in hydrogen fuel cell applications. Sealing that works reliably safeguards gas leakage while simultaneously ensuring system safety and consistent gas flow in all operating conditions. Hydrogen permeation poses a rather formidable hurdle, largely because of hydrogen's small molecular dimensions that diffuse readily through numerous commonplace materials. Viton, along with other fluoroelastomers, displays markedly superior resistance to hydrogen diffusion relative to other materials under different conditions. Viton resists oxidation pretty strongly under thermal stress and experiences minimal chemical degradation. Viton O-rings get utilized pretty frequently at really crucial junctures, like between the outlet of the electric pre-heater along with manifold supplying hydrogen. O-rings effectively seal under wildly fluctuating temperatures, amid intense mechanical vibrations, and lengthy exposure to corrosive gases and scorching heat. Their elastic properties enable absorption of physical stress rather readily, while chemical and thermal stability make them especially beneficial in fuel cell designs for automotive applications. Advanced fuel cell methods work extraordinarily well in harsh conditions because of high-temperature elastomers, which play a crucial part in performance and durability.

3.7.8 Glass Fiber and Ceramic Fiber Insulation

Insulation materials, such as glass fiber and ceramic fiber, function primarily as thermal wrapping around heaters in hydrogen fuel cell systems. Fundamental materials minimize heat loss, thereby boosting energy efficiency and stabilizing thermal conditions during system start-up and constant operation. Ceramic fibers exhibit amazing resistance to absorbing heat frequently between 600°C as well as 1000°C, and their light and adaptable framework makes it possible to insulate irregular components. Resistive heaters are usually encased in glass fiber jackets in reasonably heated PEMFC scenarios in gas pre-heating devices. Insulation proves especially advantageous in far-flung off-grid configurations such as telecom infrastructure backup power supplies, where consistent internal temperatures are crucial. These materials shield nearby parts from heat damage efficiently and help with quicker cold starts by retaining residual warmth inside system boundaries. Cost-effectiveness and straightforward installation reinforce widespread adoption throughout stationary fuel cell applications and different portable ones, rather effectively nowadays.

3.7.9 Polymers with Thermal Coatings

PTFE-coated composites and polymers indicate a considerable leap ahead in engineering low-temperature methods, particularly within hydrogen fuel cell tech. Such materials are usually utilized extensively in fabricating heat exchangers, air ducts, and control valve housings, particularly under strict compactness and weight restrictions. Their role becomes super vital in sub-zero environments where condensation and ice formation seriously compromise start-up performance and long-term fuel cell reliability. These days, numerous cases are considerably helped by PTFE's inherent hydrophobic tendency in helping relieve such issues under different circumstances. PTFE coatings help avoid condensation and subsequent freezing of water vapor inside reactant gas streams, thus protecting the integrity of inner surfaces during cold starts. PTFE-lined flow channels get integrated deeply into air pre-heater designs for portable fuel cell setups. Integration supports consistent airflow and safeguards components from risks posed by thermal stress and ice inside a moisture-resistant internal environment. PTFE's remarkable chemical stability, as well as its severe thermal durability, considerably enhances suitability for really demanding applications under harsh conditions. It remarkably endures reasonably high temperatures and oxygen exposure without degradation taking place, even under some thermal cycling conditions. Successful utilization of PTFE in these contexts underscores the wider value of superior polymer coatings in strengthening resilience inside cold-weather hydrogen systems operationally (Table 3.2).

TABLE 3.2
Review of Materials to Be Utilized in the Gas Pre-Heating Systems in the Hydrogen Fuel Cells, Including Their Applications, Winsomeness, and the Viterbi to be Anticipated in the Fuel Cell Thermal Administration

Type of Material	Application	Advantage	Primary Usage
Stainless Steel (e.g., SS316)	Heat exchanger tubes, heater casings, gas flow channels	Corrosion resistance, good mechanical strength	Structural framework and piping for pre-heaters
Inconel (e.g., Inconel 625)	Catalytic heaters, high-temperature combustion zones	High-temperature resistance, hydrogen embrittlement immunity	Burner housings and liners for hydrogen combustion
Aluminum Alloys	Microchannel heat exchangers, compact heater bodies	Lightweight, high thermal conductivity	Automotive and portable heat exchangers

(*Continued*)

TABLE 3.2
(Continued)

Type of Material	Application	Advantage	Primary Usage
Ceramic Composites (e.g., SiC, Alumina)	Heating elements, insulation blocks	Excellent thermal stability and chemical inertness	Electric heaters and thermal insulation modules
Graphite/Carbon Composites	Heat spreaders, gaskets, conductive heater parts	High thermal conductivity, inert to gases	Compact heater cores and thermal regulation plates
Aerogels (e.g., Silica Aerogel)	External insulation for heater modules	Ultra-low thermal conductivity, lightweight	Insulating heater units to reduce heat loss
Glass/Ceramic Fiber	Wrapping of heaters and duct insulation	Flexible, cost-effective heat insulation	Mid-temperature insulation in fixed or mobile systems
High-Temperature Elastomers (e.g., Viton®, Kalrez®)	Seals and gaskets in gas interfaces	Sealing flexibility at elevated temperatures	Gas-tight connections between the heater and the fuel cell stack
PTFE-Coated Polymers	Air ducts, control valve linings, low-temperature channels	Hydrophobic, prevents condensation and icing	Cold start air flow paths and moisture control
Coated Aluminum (with anti-corrosive layer)	Heat exchanger fins and casings	Lightweight, corrosion-resistant after coating	Compact air-side heating elements
Copper (with coating)	Localized heat distribution or embedded electric heaters	High thermal conductivity	Embedded heater wires or contact pads

3.8 CURRENT STATE OF IMPLEMENTATION AND PROGRESS IN PRE-HEATING THE REACTANT GASES

The use of gas pre-heating systems in hydrogen fuel cells has made great strides over the past decade. This progress stems from the growing need for reliable cold start performance and better system efficiency. Today, most commercial and prototype PEMFC systems, especially in automotive and stationary settings, include some type of gas pre-heating. This can either be a separate subsystem or part of the overall thermal management system. In automotive fuel cells, such as those found in the Toyota Mirai and Hyundai Nexo, compact electric heaters and heat exchangers help pre-heat the hydrogen and air during cold starts. These systems use waste heat from the fuel cell stack or coolant loop when the system is running. They switch to electric or catalytic heating during start-up to quickly reach operational temperatures. For instance, Toyota's system has an integrated thermal loop with quick-response heaters that reduce cold start time to less than 60 s, even in temperatures as low as −20°C.

In stationary fuel cells, like those used in Bloom Energy's SOFCs, gas pre-heating is more crucial due to the high operating temperatures of 600–1000°C. Here, regenerative heat exchangers capture exhaust heat to raise the temperature of incoming fuel and air, which greatly improves system efficiency and lowers external energy usage. Research test beds, developed by the US Department of Energy (DOE) and German Aerospace Center (DLR), have shown hybrid heating systems that combine thermoelectric generators, infrared heaters, and smart control units to optimize energy use during the pre-heating phase. These setups have achieved cold start capability at temperatures as low as −30°C while reducing energy consumption. Advancements in materials science, such as ceramic-coated microchannel heat exchangers and additively manufactured components, have created lighter, more efficient, and corrosion-resistant pre-heating modules. At the same time, AI-driven control systems are being tested to adjust pre-heating profiles based on real-time environmental conditions, fuel cell status, and historical performance (Table 3.3).

3.9 CONCLUSIONS

Pre-heating reactant gases is crucial for improving the cold start performance, operational reliability, and energy efficiency of hydrogen fuel cell systems in various applications. By warming the hydrogen and oxidant gases before they enter the fuel cell stack, this approach helps reduce problems such as slow electrochemical reactions, membrane dehydration, and ice formation during start-up in freezing temperatures. The chapter explores the basic thermodynamics and discusses different pre-heating methods, including resistive, catalytic, and regenerative systems, as well as their mixed versions, all tailored to meet specific operational needs. It also highlights the role of smart control systems, new materials, and heat recovery methods in optimizing energy use while ensuring fast and safe start-ups. Real-world examples in automotive, aerospace, stationary power, and onboard reforming systems show the need for and flexibility of pre-heating solutions. As hydrogen technologies continue to grow and adapt in colder climates and critical applications, ongoing innovation in pre-heating system design, material choice, and thermal management will be vital for supporting efficient and reliable fuel cell deployment.

TABLE 3.3
Implementation Status of Pre-Heating of Reactant Gases in Hydrogen Fuel Cell Systems

Application Sector	Example System/Company	Pre-Heating Technology Used	Implementation Status	Remarks
Automotive FCVs	Toyota Mirai, Hyundai Nexo	Electric resistance heaters, coolant loop integration	Fully Commercialized	Cold start achieved in <60 sec at −20°C; integrated into thermal system
Stationary Power Systems	Bloom Energy (SOFC)	Regenerative ceramic/metal heat exchangers	Fully Commercialized	Essential due to 600–1000°C stack operation; high energy recovery
Public Transit (Buses)	NREL/DOE Fuel Cell Bus Trials	Electric pre-heaters + insulated gas lines	Demonstration/ Deployed	Successful trials at −25°C; scalable for fleet use
Portable Power Units	Intelligent Energy, Plug Power	Miniature electric heaters, PCM-based heating	Commercial/Prototype	Used in military-grade and off-grid systems; compact, low power
Aerospace	DLR, NASA FC Systems	Infrared heaters, microchannel exchangers, smart controls	Experimental/ Prototype	Designed for high-altitude, sub-zero operations; weight optimization focus
Rail Transport	Alstom Coradia iLint (Fuel Cell Train)	Pre-heated air intake using heat exchangers	Commercial/Partial	Pre-heating integrated with HVAC/exhaust recovery; still evolving
Material Handling (Forklifts)	Plug Power GenDrive	Basic resistive pre-heaters	Commercial	Cold weather kits available; successful in refrigerated warehouses
Research Demonstrators	INSPIRE (EU Project), DOE Labs	Hybrid pre-heaters, AI-based control systems	Advanced R&D/Pilot Scale	Achieved rapid start-up at −30°C with adaptive thermal control
Marine Applications	Ballard Fuel Cell Marine Modules	Waste heat recovery, electric heating	Emerging/Prototype	Integration with marine HVAC and propulsion systems under testing
Backup Power Systems	AFC Energy, SFC Energy	Resistive + catalytic heaters, waste heat integration	Commercial/Industrial Use	Designed for off-grid and telecom backup in cold climates

REFERENCES

Amamou, A. A., Kelouwani, S., Boulon, L., & Agbossou, K. (2016). A comprehensive review of solutions and strategies for cold start of automotive proton exchange membrane fuel cells. *IEEE Access*, 4, 4989–5002.

Chang, H., Yang, Z., & Tu, Z. (2024). Experimental study on the cold-start performance of a gas heating assisted air-cooled proton exchange membrane fuel cell stack. *Renewable Energy*, 234, 121224. https://doi.org/10.1016/j.renene.2024.121224

Dieterich, M., Bürger, I., & Linder, M. (2017). Open and closed metal hydride system for high thermal power applications: Preheating vehicle components. *International Journal of Hydrogen Energy*, 42, 11469–11481.

Egeland-Eriksen, T., Hajizadeh, A., & Sartori, S. (2021). Hydrogen-based systems for integration of renewable energy in power systems: Achievements and perspectives. *International Journal of Hydrogen Energy*, 46, 31963–31983.

Gao, J., Zhou, S., Fan, L., Zhang, G., Jiang, Y., Shen, W., & Zhai, S. (2024). Innovative design of solid-state hydrogen storage and proton exchange membrane fuel cell coupling system with enhanced cold start control strategy. *Applied Sciences*. https://doi.org/10.3390/app14104068

Gießgen, T., & Jahnke, T. (2023). Assisted cold start of a PEMFC with a thermochemical preheater: A numerical study. *Applied Energy*, 331. https://doi.org/10.1016/j.apenergy.2022.120387

Guan, D., Wang, F., Zhang, X., Dou, W., & Sun, Y. (2024). Comprehensive study on catalytic coating tubular reactor with electromagnetic induction heating for hydrogen production through methanol steam reforming. *International Journal of Hydrogen Energy*, 50, 1–17. https://doi.org/10.1016/j.ijhydene.2023.07.316

Ivanova, N. A., Baranov, I. E., Kalinnikov, A. A., Mensharapov, R. M., Spasov, D. D., Sinyakov, M. V., Nikolaev, I. I., Ostrovsky, S. V., Grigoriev, S. A., & Fateev, V. N. (2023). Cold start of proton exchange membrane fuel cell using build-in catalytic heater. *Journal of Physics: Conference Series*, 2723. https://doi.org/10.1088/1742-6596/2723/1/012004

Lagrandeur, J., Uzundurukan, A., Mansour, A., Poncet, S., & Bénard, P. (2024). On the benefit of integrating vortex tubes in PEMFC system for preheating hydrogen in FCEV technologies. *International Journal of Hydrogen Energy*, 52, 1141–1152. https://doi.org/10.1016/j.ijhydene.2022.11.246

Lin, C., Kong, W., Tian, Y., Mao, Y., Zhou, E., Shao, Q., Wu, N., Liu, J., Yu, X., & Huang, S. (2023). Sandwich self-heating structure-based lithium-ion battery system and its application in the fuel cell bus for Beijing Winter Olympic Games. *Energy Conversion and Management*, 284, 116977. https://doi.org/10.1016/j.enconman.2023.116977

Luo, M., Zhang, J., Zhang, C., Chin, C. S., Ran, H., Fan, M., Du, K., & Shuai, Q. (2022). Cold start investigation of fuel cell vehicles with coolant preheating strategy. *Applied Thermal Engineering*, 201, 117816. https://doi.org/10.1016/j.applthermaleng.2021.117816

Lv, J., Shi, J., Jun, L., Mao, M., & Kong, X. (2024). Effect of preheating on extending lean extinction limit of ammonia/air combustion in a two-section porous burner. *Journal of the Energy Institute*. https://doi.org/10.1016/j.joei.2024.101708

Melnik, D., Bürger, I., Mitzel, J., Käß, J., Sarkezi-Selsky, P., Jahnke, T., & Knöri, T. (2024). Energy efficient cold start of a polymer electrolyte membrane fuel cell coupled to a thermochemical metal hydride preheater. *Applied Energy*, 359, 122585.

Moghaddam, M. S., Bahari, A., Houshani, M., Jafari, A., & Tapeh, S. M. T. (2024). A review on progress in the field of conditioning of polymer fuel cell stacks. *Journal of Power Sources* 621, 235300.

Nalluraya, N. K., Köhne, H., Köhne, S., & Konrad, M. (2011). Feasibility study of auto thermal reforming of biogas for HT PEM fuel cell applications. *International research of Brenaro Winterschool* (pp. 117–133). Springer.

Palma, V., Ricca, A., & Ciambelli, P. (2013). Fuel cell feed system based on H2 production by a compact multi-fuel catalytic ATR reactor. *International Journal of Hydrogen Energy*, 38, 406–416.

Paramanik, B., & Das, D. (2025). Accelerating nanocrystalline-diamond nucleation in the low-temperature MW-CVD growth of diamond-like carbon films via reacting-gas pre-heating. *Journal of Physics and Chemistry of Solids*. https://doi.org/10.1016/j.jpcs.2025.112554

Sinha, A. A., Ansari, M. Z., Shukla, A. K., & Choudhary, T. (2023). Comprehensive review on integration strategies and numerical modeling of fuel cell hybrid system for power & heat production. *International Journal of Hydrogen Energy*, 48, 33669–33704.

Zhan, Z., Yuan, C., Hu, Z., Wang, H., Sui, P. C., Djilali, N., & Pan, M. (2018). Experimental study on different preheating methods for the cold-start of PEMFC stacks. *Energy*, 162, 1029–1040.

Zhu, W., Wang, Y., Gao, S., He, F., Li, C., & Wang, Y. (2024). Study on cold start of fuel cell stack combined with liquid heating and fuel cell self-starting. *Case Studies in Thermal Engineering*, 60, 104814. https://doi.org/10.1016/j.csite.2024.104814

4 Electric Resistance Heating

4.1 INTRODUCTION

Electric resistance heating forms an important thermal management technique that follows the concept of Joule heating, where electrical energy is converted into heat through the flow of current through a resistive element. The simplicity of this mechanism has led to its wide application in industry to many processes as a result of its easier fabrication, fast response to thermal fluctuations, and low energy losses. In the field of hydrogen fuel cell technologies, and especially Proton Exchange Membrane Fuel Cells (PEMFCs), the electric resistance heating technology is critical to the development of cold start operation, especially in geographic areas where ambient temperatures fall below freezing. The potential effects are that under these circumstances, the transformation of water produced in electrochemical reactions into ice can present physical barriers to performance, as gas flow channels become blocked, the proton-conducting membrane becomes dehydrated, and the internal ohmic resistance is increased, and as a result of expansion caused by ice formation, mechanical damage can occur. To reduce these effects, a strategy of having strategically placed electric resistance heaters in key parts of the fuel cell system was adopted, such as the bipolar plates, flow fields, membrane electrode assemblies (MEAs), and humidification units. These heaters are sized to increase the stack temperature to working temperatures – generally between 60°C and 80°C – so that the protons conduct properly and the reacting gas pathways through the stack remain unfrozen. Many heater designs, including thin-film, foil-type, ceramic-based, and flexible polymer, are commonly used depending on the requirements of specific systems about spatial requirements, thermal uniformity, and power budget. These heating systems can be powered by auxiliary batteries on board or by utilizing a portion of the fuel cell's electrical output itself during initial operation. Moreover, sophisticated electronic control units control the real-time output of the heater to achieve optimum temperatures without localized overheating. Resistance heating can be favorably compared with other cold start approaches, such as catalytic heating or phase change materials (PCMs), in terms of direct controllability, modular architecture, and quick response time, which is why it is a commercially viable and widely implemented solution. However, difficulty in observing uniform temperature, reducing energy losses, and balancing power consumption during start-up without loss of system efficiency are still problems. Further research is still ongoing, aiming at optimization of material and control content of resistance heating systems, conductivity and energy requirements, and integration of advanced active sensing mechanisms to include dynamic thermal control. Electric resistance heating is one such revolutionary technology that contributes immensely toward enhancing the reliability, safety, and repeatability of hydrogen fuel cell start-ups with cold climate applications.

DOI: 10.1201/9781003657866-4

4.2 TYPES OF ELECTRIC RESISTANCE HEATING SYSTEMS

Electric resistance heating systems used in hydrogen fuel cell cold start applications may be categorized into broad categories depending on structural configuration, material construction, and integration pattern. They transfer electrical energy to heat through Joule heating and are designed to satisfy the peculiar needs of the polymer electrolyte membrane fuel cells (PEMFCs) used under low-temperature conditions.

4.2.1 Thin-Film Heaters

Thin-film heaters are often preferred because of their light weight, small dimensions, and ability to deposit on bipolar plates or membranes, enabling good localized heating with little thermal inertia. Thin-film heaters are becoming much more widely used to fight cold starting of a hydrogen fuel cell, particularly during cool temperatures. They offer quick, local heating that enables catching PEMFCs into quicker, more effective cool starts. The fuel cell stacks are heated by the electronic films known as thin film heaters (electrothermic films), which melt the inside ice and quickly warm the cell. The ice melts rapidly at a minimum period of 36.5 s, which operates even at extremely low temperatures, guaranteeing a high rate of cool start-up. The smaller-pored films cause ice melting quickly as compared with larger-pored films, meaning that film infrastructure affects performance.

Jiang et al. have undertaken a comprehensive numerical study to analyze the task of quick cold start-up within PEFFCs, crucial for their reliability in sub-zero environments, particularly in automotive applications. The authors created a two-phase, unsteady-state dynamic model of a 40-cell PEFFC stack to precisely reflect the ice melting behavior inside the stack during start-up, utilizing solidification/melting and multiphase flow mechanics to achieve this objective. A couple of distinct electric heating methods were examined. The proton exchange membrane (PEM) itself is heated directly, while an electrothermal film (ETF) is connected between the Gas Diffusion Layer (GDL) and the flow channel for heating. On the ANSYS Fluent software, both techniques were simulated, and the results confirmed that either technique can lead to effective and fast cold start-up. The complete melting time inside the stack was calculated to be 32.5 s with PEM heating as well as 36.5 s with ETF heating, showing the potential of a quick cool start-up. To further improve the heating efficiency, Jiang et al. investigated the effect of various ETF structural styles by simulating ETFs with different pore diameters – 0.3 mm, 0.6 mm, along 1.2 mm – while keeping a continuous porosity of 60%. Scientists discovered that scaled-down pores led to improved heat transfer and much faster melting of ice, with the smallest pore size (0.3 mm) melting entirely in 44 s compared to 48 s for 1.2 mm. The research learned that the inner cells of the stack melted before the outermost cells during the cold start process, because of convective heat losses at its boundaries. The authors examined temperature distributions, melting dynamics, and structural thermal transfer to show that ETF and PEM heating methods might satisfy the rigid cold start requirements of automotive standards (Jiang et al., 2020).

Park et al. presented a crumpled multi-walled carbon nanotube (MWCNT) thin film heater with enhanced electro-thermal characteristics for higher sensitivity hydrogen gas sensing applications. The authors created heaters that utilized an easy spray coating strategy and thermal shrinkage of polystyrene (PS) substrates to induce crumpling in the CNT films. The density of CNT junctions was considerably boosted by the crumpling effect, bringing about increased Joule heating efficiency and running conditions at reduced input voltages than those obtained from as-deposited CNT films on PET substrates. The study utilized the powerful heating ability to investigate just how crumpled CNT heaters could enhance hydrogen sensing performance, especially desorption and sensitivity. The authors examined the heaters' temperature profiles and temperature coefficient of resistance (TCR) and discovered that both kinds possessed a linear TCR behavior up to 100°C. The crumpled CNT heaters attained greater temperatures at equivalent voltages because of decreased heat loss and improved junction interconnectivity, as they discovered. This heating feature additionally permitted self-heating throughout sensing, assisting hydrogen desorption from the CNT surface and decreasing sensor hysteresis as well as resistance drift – crucial issues in regular CNT-based gas sensors. Park et al. discovered that the crumpled CNT sensors showed a lot more severe changes in electrical resistance upon hydrogen exposure than their non-shrunk counterparts because of their bigger surface area-to-volume ratio in experimental tests. The research further established that higher operating voltages and, therefore, temperatures, enhanced sensitivity and recovery, further proving the usefulness of self-heating. The study produced a framework that could double up as a heater and a gas sensor, resulting in a cost-effective, scalable, and energy-efficient platform for high-performance hydrogen detection (Park et al., 2019).

4.2.2 Foil Heaters

Flexible and ideal for integration along gas diffusion levels or to flow field plates are foil heaters are produced from metallic resistive alloys, including nichrome. The coolant loop incorporates foil heaters, and they warm a circulating fluid (like ethanol) and transport heat to the fuel cell stack. Even at temperatures close to −30°C, this technique can successfully increase the cell temperature over freezing. The system can rise to 0°C in as few as 12 s at −10°C, though lengthier heating times are needed for colder conditions (up to 53 s at −30°C) – the length of the heater as well as pump operation is essential. The maximum temperature is enhanced by longer heater operation, while pump operation guarantees a more even distribution of temperature and a steady current profile. Nevertheless, timing the pump too soon could cause failed starts or unstable voltage and current (Topcu et al., 2022).

4.2.3 Ceramic Resistance Heaters

This kind of heater features durability and high temperature stability, and it is thus perfect for pre-heating peripheral components like humidifiers or water separators. Duan et al. have completed research on protonic ceramic fuel cells (PCFC) that offer considerable details, which are crucial to moving forward with hydrogen fuel cell cold start solutions. PCFCs exhibit good performance at intermediate temperatures (500–600°C), leading to a smaller thermal gap from background conditions, enabling quicker and much more energy-efficient initialization. The simple fact that they can self-heat through exothermic electrochemistry reactions and in situ fuel reforming removes outside heating methods, which is a crucial function in cold climates where auxiliary energy usage must be lessened. Particularly, the in situ exsolution of nickel nanoclusters in earlier operation encourages catalytic activity and velocities hydrogen oxidation and internal heat creation, crucial for fuel cell function under cold start conditions. Additionally, their demonstrated resistance to cooking and sulfur poisoning guarantees dependable operation even during fuel-rich and transient working conditions typical of start-up phases. The BZY-based electrolyte integrated into these cells encourages water-mediated carbon cleaning by area hydroxylation, a method which could help avoid ice buildup and degradation during repeated freeze–thaw cycles. The cells display superb thermal cycling durability and regain over 99.5% of their original functionality after just 32 rapid heating and cooling cycles, making them perfect for applications requiring regular starts in cool temperatures. PCFCs possess fuel flexibility, enabling them to utilize hydrogen, ammonia, hydrocarbons, and methanol immediately, which may help hybrid cold start techniques employing much less reactive or thermally useful fuels before evolving to hydrogen. These features collectively underscore the significance of PCFC technologies in enabling as well as improving the creation of strong, fuel-tolerant, and thermal-resilient hydrogen fuel cell cold start systems (Duan et al., 2018).

Liu et al. have developed a very efficient PCFC by utilizing a new anode material, Ni $Ba_{0.96}$ ($C_{0.66}Zr_{0.1}Y_{0.2}Ni_{0.05}$) O_3 (Ni-BCZNY), which utilizes exsolved nickel nanoparticles in situ to boost electrochemical activity at intermediate to low heat. The authors managed to attain uniform exsolution of Ni nanoparticles in the reduction of hydrogen by utilizing A-site deficiency and controlled Nickel doping in the perovskite structural configuration, leading to improved catalytic activity, hydrogen dissociation, and charge transfer kinetics. The ensuing PCFC offered great performance, obtaining a good power density of 912 $mWcm^{-2}$ at 700°C and keeping a measurable output while at 250°C. The decrease in polarization resistance to 0.04 cm^2 and reduced activation power was confirmed by electrochemical impedance analysis, and durability tests demonstrated long-term operational stability with little degradation over time. The low-temperature functionality and enhanced self-heating capability demonstrated in this research support fast start-up under sub-zero conditions, making these findings crucial for hydrogen fuel cell cold start solutions. Exsolved Ni nanoparticles work as energetic catalytic sites at lower temperatures and can accelerate hydrogen oxidation and internal heat production, therefore lowering external heating demands. The effective thermal cycling stability and structural integrity of the anode electrolyte interface enable the cell's capability to work in cool start environments, making this PCFC design a practical path toward far more robust as well as efficient hydrogen fuel cell technologies that could operate reliably in cold environments (Liu et al., 2020b).

To provide a compact, one-step technique for creating high-purity hydrogen, Wang et al. produced and also demonstrated large-area protonic ceramic cells working with a NiO BZCY combination for electrochemical hydrogen pumping (EHP). The authors obtained considerable mechanical strength enhancements while maintaining good electrochemical performance by constructing twelve 12 cm^2 planar cells using thick (350 μm) and low porosity (9%) electrodes. The cells experienced poor polarization resistance, achieving 0.12 Ω cm² at 800°C and producing a hydrogen flux of 8.7 mL $cm^{-2}min^{-1}$ at 650°C with 90% Faradaic efficiency. The study utilized impedance spectroscopy and distribution in relaxation time (DRT) analysis to pinpoint dissociative hydrogen adsorption and diffusion as rate-limiting mechanisms, which varied with temperature. The cells additionally demonstrated good thermal and structural stability during dual-atmosphere and durability tests, with minor Ni nanoparticle migration and coarsening, each without substantial degradation, indicating their reliability in real-life operating conditions. The results have substantial importance for hydrogen fuel cell cold start solutions. The PCCs had been initially developed for hydrogen purification; however, their effectiveness at intermediate temperatures makes them a good way of internal heat generation during the cold start of a fuel cell. The EHP mechanism not only helps with hydrogen transport and concentration but also inherently creates heat, which could be leveraged to pre-warm elements or gas streams in PEMFCs during start-up under sub-zero conditions. By examining gas diffusion, catalytic activity, and material behavior under electrochemical and thermal stresses, the research can also help refine auxiliary heating elements or integrated hydrogen management systems in fuel cell stacks. The PCCs possess structural integrity and thermal cycling stability, which make them perfect for cold start applications where dependability, quick response, and durability are crucial. The technologies as well as observations presented by Wang et al. indicate a possible path to enhance the resilience as well as operability of hydrogen fuel cells under cold conditions (Wang et al., 2022) (Figure 4.1).

4.2.4 PTC (Positive Temperature Coefficient) Heaters

The natural safety of PTC heaters is acquired through their self-regulating thermal behavior, which reduces power draw as Temperature increases. Heat is produced by passing an electric current through materials whose resistance decreases with temperature, which is inherently safe and temperature-controlled. In FCVs, PTC heaters are generally activated once the car begins and the outdoor temperature is below a

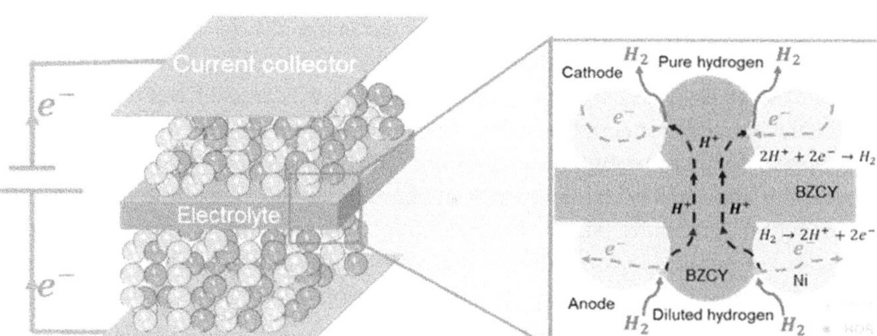

FIGURE 4.1 Electrochemical hydrogen pumping (EHP) employing protonic ceramic cells: working concept. HOR and HER stand for hydrogen evolution reaction (HER) and hydrogen oxidation reaction, respectively (Wang et al., 2022).

set threshold, using electricity from the fuel cell stack to warm the cabin or even assist in cool starts. Cao et al. (2024) studied precisely how self-regulating qualities helped minimize thermal overshoot risks in stack preconditioning. Cao et al. proposed an innovative and efficient cold start approach for PEMFCs, which utilized a genetic algorithm (GA) to tweak control parameters. The authors created as well as validated a detailed simulation model of a 114-kW PEMFC system to deal with the issue of constructing PEMFCs at incredibly low temperatures, such as 40°C, where ice formation in the catalyst layer can obstruct reactions and damage parts. Their multi-stage heating approach dynamically adjusted stack current and PTC heater power, dependent upon stack temperature intervals. The optimal weighting coefficient (obtained by GA) for the strategy with a GA derived optimum wattage of 0.7 led to a start-up time reduction of 32.39% along with a hydrogen usage of 36.17% compared to passive methods, while consuming 17.54% less heater energy than constant full power heating. This strategy led to fast cold starts of 77.6 s, increased energy efficiency, and outperformed other standardized cold start techniques. The work represents a tremendous improvement in hydrogen fuel cell cold start solutions, as it shows just how coordinated external and internal heating control may efficiently fight ice issues and decrease energy demands. The capability to alter current and heater power in response to heat guarantees quick, uniform warming of the stack, stopping ice accumulation and encouraging system durability. The study further shows that controlled internal heat creation is more efficient than depending on just external heaters. The GA-based optimization offers a dependable framework for varying cold start strategies, making it particularly helpful for automotive fuel cell systems. The study encourages the development of intelligent thermal management systems targeted at next-generation hydrogen fuel cell automobiles running in chilly environments (Cao et al., 2024).

Yang et al. created a durable thin film Positive temperature coefficient (PTC) material which has a high resistance temperature Coefficient ($2.8°C^{-1}$) and was built to guarantee self-regulating and precise Temperature control. The group created a lightweight PTC composite which has two unique Curie temperatures (30°C and 38°C), high thermal cycling repeatability, along with an impressive PTC intensity of 6.5 by combining ethylene vinyl acetate propolymer with dodecanoic acid and carbon nanotubes. The experimental setup confirmed that the material kept the temperature associated with a heated object (an aluminum block) with fluctuating ambient factors and various voltage inputs. A specific theoretical model was created and tested, demonstrating that this PTC material-based heating system can maintain consistent temperatures even under extreme environmental conditions, such as changes in environmental temperature and wind speed, allowing temperature swings of as little as 0.13–2.7°C depending on conditions. These results tend to be promising for hydrogen fuel cell cold start solutions. Ice formation at sub-zero temperatures inhibits cool starts in PEMFCs, requiring fast and energy-efficient thermal management methods. This research sought to produce brand new PTC materials that are self-regulating, compact, as well as low-energy heating solutions which could be directly incorporated into a fuel cell stack to preserve part temperatures above freezing. The high resistance temperature coefficient assures the heater normally restricts its power output as the ideal temperature approaches the desired range, preventing overheating and lowering

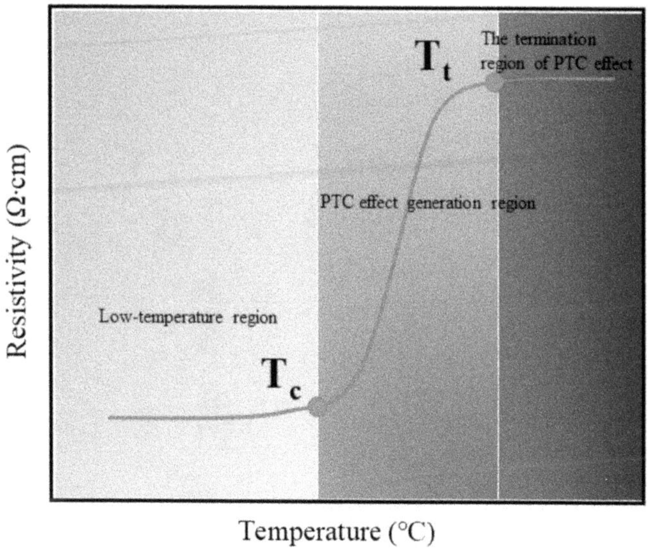

FIGURE 4.2 The resistance temperature qualities of PTC materials (Yang et al., 2024).

the need for extra sensors or controllers. The material's superb thermal cycling stability assures long-term reliability throughout repeated start–stop cycles that are typical in portable and automotive applications. This article offers a solid argument in favor of lightweight, effective, and durable electric heating technologies that meet the cold start requirements of hydrogen fuel cells (Yang et al., 2024) (Figure 4.2).

The resistivity profile associated with a PTC material across three different temperature ranges to understand its self-regulating heating attributes is shown in Figure 4.3. This particular material operates like a typical conductor in the lower temperature area with continuous and minimal resistivity. The resistivity begins to rise sharply as the temperature approaches the Curie temperature (Ta), and that is the PTC effect generation region. A very slight rise in temperature leads to a significant degree of resistance, enabling the material to restrict current flow and avoid overheating. The resistivity eventually stabilizes outside of the termination temperature (Tt) in the termination area of the PTC effects, where extra temperature rises just an impact on resistivity. PTC materials with this behavior are perfect for thermal management applications like cold start solutions in hydrogen fuel cells, where heat must be maintained within efficient and safe operating limits (Yang et al., 2024).

4.2.5 Carbon-Based Resistance Heaters

Carbon-based resistance heaters are becoming revolutionary parts in hydrogen fuel cell systems, particularly for enhancing compactness and efficiency. Carbon materials are utilized in these heaters to offer localized, direct heating, which may enhance fuel reforming and overall system performance. High-efficiency resistive elements made from carbon, like carbon nanotube films and graphene-infused composites,

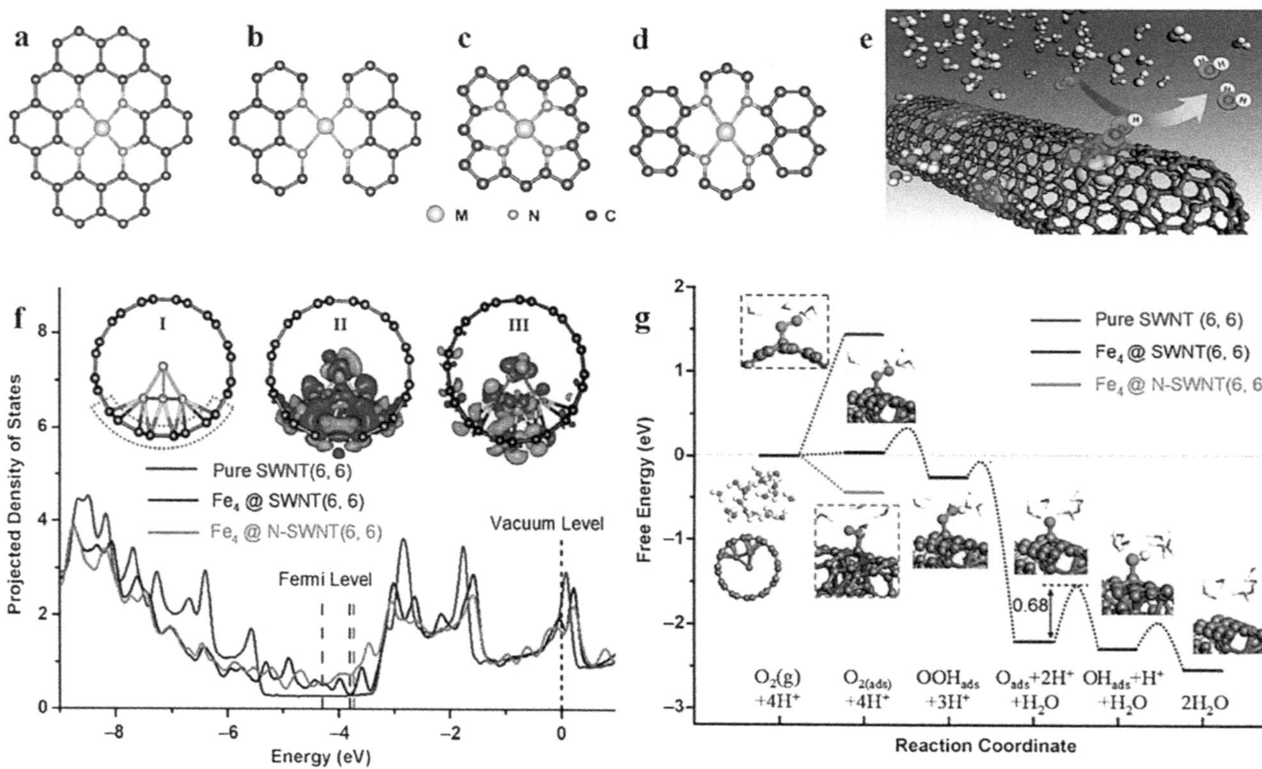

FIGURE 4.3 (a–d) Molecular structures of several likely active sites of M-N-C catalysts. (e) Schematic diagram of the ORR process happening on the surface of the Fe4 SWNT catalyst. (f) Projected DOS of the p orbital of the C atom bonded to Fe4 vacuum amount is aligned to zero eV. (g) ORR free energy level diagram on Fe in the three materials (Chen et al., 2023).

are becoming more and more being used as higher thermal conductivity as well as electric performance alternatives. Yang et al. investigated the integration of cylindrical heat pipes (CHPs) into an altered reversible MRPEMFC to enhance both thermal management and electrochemical hydrogen storage capability. The experiment incorporated a laboratory-scale MRPEMFC with a self-standing activated carbon electrode and CHPs incorporated in the hydrogen and oxygen bipolar end plates. A pair of operating modes has been evaluated: one with CHPs (mode 1) and another without (mode 2). The hydrogen storage capability in mode 1 increased by 40% to 1.52 wt% in comparison with 0.91 wt% in mode 2. The main reason for this improvement is the decrease in operating temperature on the hydrogen side by 24% (12°C) throughout the charging process, which enabled better hydrogen adsorption into the active carbon pores. Mode 1 had somewhat decreased current output during discharge but supplied a greater initial voltage and much better voltage retention because of more stable hydrogen bonding and thermal circumstances. The results have considerable importance for hydrogen fuel cell cold start solutions because of the significance of thermal regulation for operation in sub-zero environments. Among the primary challenges during cold starts is stopping ice formation in the membrane and maintaining conditions ideal for electrochemical reactions. CHPs offer an energy-efficient and practical option for stabilizing internal temperatures during start-up without the need for external power-consuming heating systems, through passive thermal management. By guaranteeing much better hydrogen storage and smoother release during operation, the CHP-integrated design supports both fast as well as dependable cold start performance (Singh et al., 2024).

Chen et al. conducted an extensive review of the usage of carbon-based substances as low-platinum catalyst supports and noble catalysts in polymer electrolyte fuel cells (PEFCs), concentrating on application in the oxygen reduction response at the cathode. The research significantly examined the drawbacks of standard carbon black supports, including their corrosion resistance, weak metal support interaction, as well as absence of intrinsic catalytic activity, and highlighted the advantages of superior carbon materials created using different template-induced techniques (hard, very soft, and self templates, and also combined templates). Advanced carbons, including doped graphitic structures, defective carbides, along with metal–nitrogen–carbon (M-N-C) complexes, displayed higher conductivity, a bigger surface area, enhanced corrosion resistance, and superior ORR catalytic performance. The review additionally examined synthesis mechanisms, structural optimization methods, along with electrochemical attributes which enable these materials to either support ultra-low platinum loadings or even function as viable alternatives to platinum altogether. The results are especially pertinent to hydrogen fuel cell cold start solutions, where kinetic efficiency and stability of ORRs are significant problems at lower temperatures. Ice formation and sluggish electrochemical reactions can significantly affect performance during cold starts. Highly developed carbon-based catalysts, particularly those with pore structures that are optimized for energetic site density and intrinsic conductivity, could help develop much more powerful and corrosion-resistant ORR pathways even during extreme cold conditions. Their large surface area as well as electrochemical stability support better reaction kinetics throughout start-up, lowering voltage losses and ensuring smoother cool start transitions. Additionally, their longevity throughout thermal cycling and resistance to degradation make them ideal for repeated cold start events in handheld and automotive fuel cell systems. Chen et al.'s research not only advances the development of new catalyst materials for PEFCs but also provides crucial design guidelines for enhancing the cold start performance of hydrogen fuel cell technology (Chen et al., 2023).

In Figure 4.3, the electronic and structural improvements of carbon nanomaterials, especially nitrogen-doped single-walled carbon nanotubes (N-SWCTs), are discussed, and their effects on the oxygen reduction reaction (ORR) when coupled with transition metals like iron (Fe). Panels (a) to (d) illustrate the atomic configurations of different catalyst systems: N-SWCNTs were anchored with fresh SWCNT, nitrogen-dipped carbon, and Fe, creating Fe-N-C sites. These structural modifications demonstrate just how heteroatom doping and metallic coordination considerably modify the electronic properties of the carbon matrix to boost catalytic activity. In panel (e), oxygen molecules are shown interacting with Fe-N-C-containing sites on the carbon nanotube surface, highlighting the importance of these engineered active sites in facilitating efficient ORR pathways. Panel (f) presents the projected density of states (PDOS), comparing FeN-SWCNTs, FeSWCNTs, and pure SWCNTs. The results reveal that nitrogen doping and Fe incorporation shift the Fermi level and also raise the electric density near it, indicating enhanced electric conductivity and catalytic potential. The insets further illustrate charge delocalization in FeN-SWCNT, which is beneficial for facilitating the adsorption as well as transformation of ORR intermediates. The reaction free energy diagrams for the ORR process for various materials are shown in panel (g), with Fe-N-SWCNT displaying the lowest energy obstacles all throughout the ORR path, indicating its superior catalytic efficiency. Figure 4.3 emphasizes the possibility of innovative carbon-based catalysts (e.g., Fe-N-C-modified SWCNTs) for hydrogen fuel cells. Their improved conductivity and structural stability, along with lower ORR activation energy, make them extremely ideal for low temperature conditions encountered during cold starts. These materials offer significant benefits in hydrogen fuel cell cold start solutions by enabling effective as well as strong electrochemical performances even at sub-zero temperatures, eliminating the need for platinum and facilitating the creation of durable, cost-effective, and highly useful PEMFC systems.

4.2.6 Microheaters

Microheaters are often embedded in sensors or localized zones of the MEA and are employed for quick spot heating, thereby enhancing system responsiveness. The heaters may be run straight from the vehicle battery or auxiliary energy units, and also, the more advanced systems include temperature sensors and control electronics for real-time feedback and optimized energy use. Microheaters offer localized, fast, and controllable heating within microreactors, which is crucial for initiating and maintaining high-temperature reactions like steam or even dry reforming that transform hydrocarbons into hydrogen-rich gases for fuel cell applications. Integrated microheaters offer substantial energy savings, enhance thermal efficiency, and allow for a smaller system footprint compared with traditional outside heating strategies. Their quick thermal response supports rapid start-up and dynamic operation, making them ideal for on-demand or portable hydrogen generation. Furthermore, microheaters facilitate process intensification, supporting greater throughput and compact reactor integration in micro fuel cell systems and portable electronics. Microheaters fabricated through MEMS methods are presented by Zhou et al. (2024), who embedded platinum microheaters into catalyst-coated membranes for fast cold start activation. Cui et al. designed and produced a Surface Acoustic Wave (SAW) hydrogen sensor incorporating a carbon-based Pd/Ni thin film and a microheater, aiming to considerably improve the response speed and sensitivity of hydrogen detection methods. The sensor is based on a 200 MHz SAW chip installed on a Y35° X quartz surface coated with a 50 nm thick layer of Pd/Ni alloy thin film, where nickel is added to avoid phase transition in palladium and improve structural stability. The sensor exhibited a quick response time of under 2 s, a very low detection limit (<15 ppm), minimal energy consumption (<1.35 W), and outstanding stability and humidity resistance. The microheater controls the working temperature of the delicate movie to an optimal 75°C, which speeds up hydrogen diffusion into the Pd/Ni layer, enhancing sensing properties. The extensive characterization confirmed consistent structural integrity, temperature uniformity, strong repeatability under different hydrogen concentrations as well as humidity levels, and powerful repeatability. These observations are particularly pertinent in hydrogen fuel cell cold start scenarios where quick and accurate hydrogen detection is vital at lower temperatures. Carbon-supported Pd/Ni thin films offer improved catalytic activity and thermal stability, enabling powerful real-time monitoring throughout start-up. The sensor can work effectively while below freezing because of the integrated microheater, enabling early leak detection and system readiness. The sensor is particularly ideal for automotive fuel cell systems, which call for quick response and thermal resilience during cold starts because of its low power consumption and high sensitivity (Cui et al., 2023).

Figure 4.4 demonstrates a SAW-based hydrogen sensor structure created to boost cold start performance in hydrogen fuel cells. The sensor includes several main elements – a sensitive film for hydrogen detection, a microheater to regulate localized temperature, interdigital transducers (IDTs) for producing as well as receiving aural waves, bus bars for electrical connectivity, and a protective layer all constructed on a piezoelectronic crystal substrate. During cold start conditions, when low temperatures prevent hydrogen detection and reaction kinetics, the integrated microheater guarantees dependable sensor operation by preventing frost and keeping optimum temperatures. The very sensitive film rapidly reacts to hydrogen appearance, enabling real-time monitoring and control of hydrogen flow into the fuel cell. The fuel cell start-up efficiency and dependability are improved in freezing environments by this compact, thermally handled sensor.

4.2.7 Flexible Printed Circuit (FPC) Heaters

Flexible printed circuit (FPC) heaters are gaining interest for their potential to enhance thermal management in hydrogen fuel cells, which is crucial for effective operation, particularly in cool environments. While FPC heaters are widely studied for portable and wearable electronics, their immediate application in hydrogen fuel cells is an emerging area with limited but increasing research. These heaters

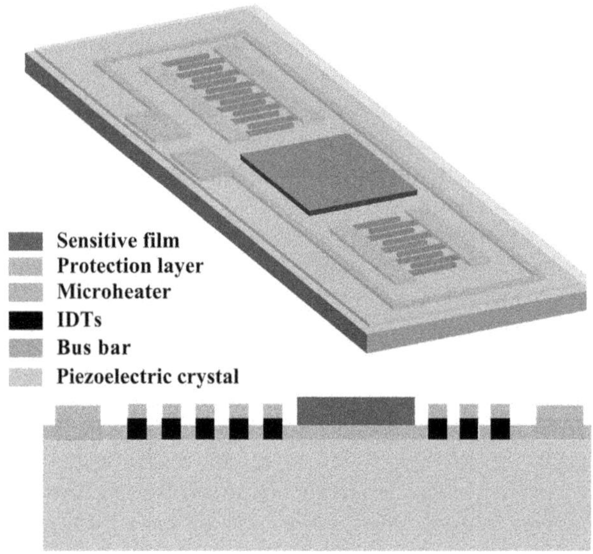

FIGURE 4.4 Representation of SAW hydrogen sensor in 3D overall and cross-sectional view (Cui et al., 2023).

are fabricated utilizing polyimide substrates, which permit seamless incorporation into curved or constrained areas inside the fuel cell stack. Liu et al. (2020b) proved their good application in complicated geometries as serpentine flow fields, enhancing thermal uniformity throughout the cell stack. Liu et al. created a cost-effective, but extremely effective, printed flexible heater (FH) based on oxidation-resistant copper–nickel rose-stem nanowires (Cu-Ni RSNWs) with the aim of providing a scalable platform for thermal management with several applications. The authors created heaters that were incredibly resistant to oxidation, thermally stable, and physically flexible, and maintained conductivity after 1000 bending cycles while heating stably up to 172.8°C with little power input, by screen printing Cu-Ni conductive ink onto substrates like paper. These characteristics directly relate to the requirements of hydrogen fuel cell cold storage, where reliable, uniform, and efficient heating is crucial to avoid freezing and facilitate system start-up in low-temperature environments. The heaters demonstrated quick thermal response and solid durability, which can be crucial for warming elements like the MEA and gas lines. Additionally, the successful use of these heaters in warming blood bags and wearable thermotherapy implies their adaptability for portable and integrated heating solutions (Liu et al., 2020a). Liu et al. provide an extensive introduction to the research on printed flexible resistive heaters (FRHs), addressing their materials, fabrication techniques, and potential uses in portable and wearable thermal management. They highlight conductive materials like conductive polymer, carbon-based nanomaterials (CNTs, graphene), along with metal-based nanowires (NWs) (such as Ag NWs and Cu-Ni NWs) and hybrid composites, which are integrated into stretchable or flexible substrates using scalable printed electronics (PE) methods, including screen printing, inkjet printing, rolling-to-roll (R2R), as well as 3D printing. The heaters possess a quick thermal response, uniform heating, mechanical durability, and minimal power consumption, making them compatible for large area, high-performance, and low-cost applications. These printed flexible heaters are vital thermal management products for hydrogen fuel cell cold storage since they prevent freezing of crucial parts like the MEA or GDLs. Their capability to quickly produce as well as distribute heat at low voltage guarantees dependable cold starts in sub-zero environments, which is crucial for both portable and vehicle-based hydrogen fuel cell systems. In addition, their bendable and lightweight design enables seamless integration into intricate fuel cell architectures without adding bulk or compromising performance, providing a practical solution to just about the most persistent issues of hydrogen fuel cell technology – cold start capabilities (Liu et al., 2021).

4.2.8 Surface Mount Heaters

Surface mount heaters are crucial components in hydrogen fuel cell systems, especially for ensuring dependable functioning in frigid environments and optimizing hydrogen release from storage. Their integration as well as performance directly influence fuel cell efficiency, start-up dependability, and total system safety. Heaters are utilized where replaceable or modular heating is desired. Each system is chosen according to its effectiveness, durability, response time, and compatibility with the fuel cell architecture. The strategic choice and positioning of these heaters considerably influence cold start efficiency, energy usage, and component longevity. Surface and integrated heaters are crucial for facilitating cold starts and effective hydrogen release in hydrogen fuel cell systems. For PEMFCs, integrated catalytic heating units may rapidly increase inner conditions from as small as –40°C to operating temperatures, guaranteeing secure start-up and preventing damage to the membrane, dehydration, and harm. Heat exchangers or heating units within metal hydride (MH) storage systems are employed to keep the ideal tank surface temperatures for healthy hydrogen discharge and uniform fuel cell operation. Several heater technologies support these functions: catalytic heaters, integrated directly into fuel cell plates, offer quick, even heating; thermochemical pre-heaters, which use metallic hydrides, offer compact, self-sufficient thermal management solutions perfect for cold climates; and heat exchangers or heat pipes regulate both cooling and heating to enhance hydrogen adsorption/desorption and maintain secure system temperatures. Current studies have emphasized hybrid configurations that combine different heater types to attain a consistent temperature distribution across the fuel cell. These varied electrical resistance heating systems collectively form the foundation of contemporary thermal management methods important for cold start solutions in hydrogen fuel cell–powered automobiles. Elkhatib et al. investigated a green, hydrogen-powered micro-combined heat and power (micro-CHP) system utilizing a PEM fuel cell for residential applications, emphasizing thermal efficiency and energy self-sufficiency. The system brings together renewable sources (solar and wind) for electrolysis of hydrogen, which runs the PEMFC to create heat and electrical energy. Their experimental setup, verified via dynamic simulations, demonstrated strong performance throughout seasonal conditions, especially under a heat-led strategy where thermal demand drives system operation. During chilly months, the fuel cell operates for extended periods, with heat efficiently kept in buffer tanks to meet space heating and hot water needs. Critical parts, including the hydrogen supply lines, the water tanks, and the PEMFCs themselves, are pre-heated by surface-mounted heaters in this particular system, especially during cold starts. The heaters ensure that the system reaches operating temperatures quickly – even in sub-zero conditions – stopping membrane dehydration, frOST formation, and efficiency losses. They enhance hydrogen flow and fuel cell response time through quick, localized heating. The authors discovered that with the best thermal management – along with surface heating units – the system can achieve 100% thermal coverage and as much as 120% hydrogen autonomy in favorable climates. Their results highlight just how integrated heating technologies improve system responsiveness and resilience. Surface-mounted heaters turn into crucial ingredients for year-round reliability and cold start readiness in hydrogen fuel cells for residential power systems (Elkhatib et al., 2024).

Pouya Barnoon developed and simulated a high-temperature, carbon-based propane steam reforming heat exchanger system for hydrogen generation for fuel cells. The cylindrical surface-mounted heating tubes are incorporated in a packed-bed reformer and also function as surface heaters used in cold starting solutions for PEMFCs. These tubes deliver localized, controllable heat to drive the endothermic response, crucial for initiating hydrogen generation, particularly under sub-zero start-up conditions. The model simultaneously examined thermal, hydrodynamic, and chemical fields, obtaining over 92% hydrogen yield and near complete propane conversion at optimum conditions. The study identified a permeability of 1×10^{-9} m^2 as optimum and also emphasized the roles of hot gasoline flow rate as well as temperature for regulating heat transfer, response kinetics, and hydrogen paper. The creation of more than 96% CO_2 together with H_2 demonstrates efficient carbon use. These findings not only advance carbon-based hydrogen reforming technologies but also offer design techniques for integrating surface heaters in fuel cell systems to ensure effective and quick cold starts (Barnoon, 2021).

4.3 MATERIALS FOR ELECTRIC RESISTANCE HEATING ELEMENTS

The selection of materials for electrical resistance heating components is vital for attaining effective, reliable, and durable functioning in hydrogen fuel cell cold starting systems, particularly in sub-zero conditions. The ideal material must have high electrical resistance, excellent thermal conductivity, oxidation resistance, mechanical strength, and stability at elevated temperatures.

4.3.1 Nichrome (Ni-Cr Alloys)

Nichrome (Ni-Cr alloys), famous due to its increased resistance (~1.10–1.50 μΩ·m) and good corrosion resistance, makes it ideal for extended heating applications. Nady et al. (2021) highlights its capability to stand up to temperatures up to 1200°C with no substantial deterioration, and that is beneficial for thermal pre-conditioning of fuel cell stacks. The authors examined new carbon-based Ni-Cr combinations, like Ni-Cl-Mo-Fe, Ni-Cl-Mo, and Ni-Cr, as potential catalysts for the HER in alkaline electrolytes, a crucial part in hydrogen fuel cell systems, and created as well as examined them in this study. The inclusion of Fe and Mo on the Ni-Cr alloy greatly enhances the catalytic efficiency, with the Ni-Cr-Mo-Fe alloy showing the greatest electrocatalytic effectiveness – attaining a reduced overpotential of 232 mV at 10 mA cm^{-2}, along with a Tafel slope of 57.7 mV dec^{-1} in 1.0 M KOH, surpassing several other non-noble catalysts. The increase in active sites and also the improvement of charge transfer kinetics are credited for the enhancement, with electrochemical impedance spectroscopies (EIS) demonstrating a reduced charge transfer resistance (Rct) and a greater electrochemical surface area (ECSA), which permits faster hydrogen generation at lower energy input. These materials hold potential in hydrogen fuel cell cold start solutions on account of their increased activity at lower overpotential and low activation energies (a range as low as 8.9 kJ mol^{-1}) that may permit quick hydrogen production and system warm-up at lower temperatures, overcoming a significant obstacle to fuel cell start-up. Consequently, the proposed Ni-Cr-Mo-Fe alloy emerges as a cost-effective and efficient option for next-generation hydrogen fuel cell technologies, particularly under challenging start-up scenarios (Nady et al., 2021).

4.3.2 Kanthal (FeCrAl Alloys)

Kanthal (FeCrAl alloys) is ideal for numerous energy applications because of its high oxidation resistance and long-lasting properties. In hydrogen fuel cells, particularly during cold starts, materials must tolerate rapid temperature changes and stop hydrogen permeation or degradation. Kanthal can maintain structural integrity at temperatures over 1400°C, according to Gehmlich et al.'s (2025) research, making it suitable for systems needing powerful thermal cycling. Kanthal variations of FeCrAl alloys possess a steady aluminum oxide layer on the surface, which stops hydrogen permeation. The usefulness of this barrier depends upon the alloy's microstructure; nanocrystalline and ultra-fine-grained forms display higher activation energies for hydrogen permeability when compared with coarse-grained samples. This indicates that FeCrAl alloys can restrict hydrogen transport, which could be applicable for hydrogen containment or safety in fuel cell environments, although not particularly for cold start solutions in fuel cells (Gehmlich et al., 2025).

4.3.3 Carbon-Based Materials

Hydrogen fuel cells are a promising, cleaner power technology; however, their efficiency is seriously restricted during cold starts, particularly in automotive applications. Carbon-based materials can make it possible to enhance different components of hydrogen fuel cells and may also provide answers to cold start difficulties. Carbon nanotube (CNT) films and graphene composite materials have lately received interest because of their lightweight, outstanding power conductivity, and flexible layout possibilities. The electrical and mechanical properties of bipolar plates are improved by carbon fibers, expanded graphite, or carbon nanotubes, which are vital for performance and durability in cold starts. Gehmlich et al. studied hydrogen permeation of FeCrAl alloys with various microstructures and also discovered that ultra-fine grained (UFG) and nanocrystalline (NC) constructions had improved hydrogen solubility as well as permeability than coarse grained samples. This behavior was primarily attributed to the enhanced grain boundary density in finer structures, which function as reversible hydrogen traps that improve diffusion and uptake. Moreover, thin aluminum oxide (Al_2O_3) surface layers formed as well as served as diffusion barriers, therefore preventing hydrogen transport. While the research concentrated on metallic systems, the concepts demonstrated – particularly, the influence of microstructure and surface chemistry – are specifically applicable to the design as well as function of carbon-based materials used in hydrogen fuel cell cold start applications. Carbon-based materials, including carbon black, graphene, and doped porous carbons, are frequently utilized as catalyst supports or electrode components in hydrogen fuel cells. Their structural characteristics – including high surface area, porous structure, and tunable defect densities – permit them to work efficiently under cold start conditions where hydrogen transportation as well as electrochemical reactions are hindered. Just as improved grain boundaries in UFG metals improve hydrogen trapping, imperfections and edge sites in carbon materials offer active adsorption areas for hydrogen, supporting reaction initiation at lower temperatures. Additionally, functionalized carbon surfaces, like oxide-covered metallic alloys, could either improve or impede hydrogen access depending on their surface chemistry. Carbon-based materials can guarantee efficient hydrogen delivery and electrochemical performance by carefully engineering these characteristics, even when ice formation or slow system kinetics complicate system operation (Table 4.1).

4.3.4 Ceramic–Metal Composites

Cermets, ceramic–metal composites, are a well-known material for advanced energy systems due to their amazing ability to maintain high temperatures and electric conductivity. In hydrogen fuel cells, especially under tough cold start or high temperature conditions, the selection of cermet materials is essential for durability and performance. The resilience of Molybdenum (Mo) cermets in high-temperature hydrogen environments was evaluated, often with yttria-stabilized zirconia (YSZ) or uranium dioxide (UO_2). These cermets demonstrate strong hydrogen compatibility, little mass loss, and good retention of inside particles, even after exposure to temperatures up to 2630 K. Microstructural degradation might occur, which includes cracking and swelling, particularly at high temperatures or after repeated thermal cycling. The Mo

TABLE 4.1
Functions of Different Carbon-Based Materials Concerning Cold Start in Hydrogen Fuel Cells

Material Type	Function in Fuel Cell	Cold Start Benefit	Citations
Graphene, CNTs	Catalyst support, BPPs	Improved conductivity, durability	Gehmlich et al. (2025)
Porous Carbon	Water management, storage	Reduced ice formation	Xia et al. (2024)
Carbon Nanofibers	Bipolar plates, electrodes	Enhanced heat transfer	Zhou et al. (2024)

matrix usually continues to be strong, but the ceramic phase (YSZ) could experience grain boundary attack and phase changes under extended hydrogen exposure (Duffin et al., 2023). The conventional anode components in solid oxides (SOFCs) and PCFCs are cermets, especially Ni-YSZ, and Ni-based alloys. They offer excellent catalytic activity for hydrogen dissociation and excellent mechanical integrity under thermal cycling. Alloying with metals like Mo or W can further enhance tolerance to carbon deposition and enhance durability in hydrogen-rich environments (Osinkin, 2023).

4.3.5 Stainless Steel

Due to its cost-effectiveness, mechanical toughness, as well as oxidation resistance, stainless steel is frequently used in hydrogen fuel cells, particularly for bipolar plates. Cold start conditions and hydrogen exposure present issues as corrosion, hydrogen embrittlement, and decreased conductivity. Recent research investigates material modifications and surface treatments to deal with these issues and enhance cold start performance. The variable microstructure, corrosion resistance, as well as electrical conductivity of stainless steel, particularly 316L, contribute considerably to the long-lasting performance of hydrogen fuel cells under cold start conditions. Superior processing methods, including Selective Laser Melting (SLM) and Surface Mechanical Rolling Treatment (SMRT), refine the microstructure and also generate thicker, much more protective passive films, drastically boosting corrosion resistance and conductivity, essential for maintaining dependable functioning in humid and cold environments. Interfacial contact resistance is decreased and current flow enhanced through surface alterations, including ion implantation (e.g., with chromium or molybdenum) and the application of conductive coating (including polyaniline/Zn-Porphyrin composites). SLM-produced 316L stainless steel likewise exhibits enhanced resistance to hydrogen-induced damage compared to conventionally processed steels, owing to its robust passive films and stable austenite phase. Cold deformation associated with a reasonable amount can enhance corrosion resistance via grain refinement and film stability, but excessive deformation leads to martensite formation and also raises susceptibility to hydrogen embrittlement. Additionally, controlling hydrogen uptake is vital because cold working or extended exposure to high-pressure hydrogen can cause brittleness and ductility loss, particularly in austenitic and duplex-grade materials. Engineering passive films, particularly thin, chromium-rich layers or stable oxide coatings on austenitic stainless steels, not only enhances corrosion resistance but also boosts electrical conductivity, both crucial for ensuring effective, quick cool start and long-term dependability of hydrogen fuel cells (R. Wang et al., 2024).

4.4 DESIGN AND CONSTRUCTION OF RESISTANCE HEATERS

Resistance heaters for hydrogen fuel cell cold start are needed for rapid and dependable thermal activation under sub-zero conditions. The heaters are designed to provide precisely controlled, localized heat to crucial parts like the MEA, the flow fields, humidifiers, and coolant systems, which are prone to freezing and performance degradation. Resistance heaters operate based on the Joule heating concept, where electrical current passing through a resistive material produces heat. Nichrome, Kanthal, stainless steel alloys, carbon composites, and printed conductive inks tend to be frequently used materials, selected for their thermal efficiency, durability, and compatibility with the fuel cell environment. Usually, the heaters tend to be thin films, supple foils, or embedded ceramic elements, which can be easily incorporated into fuel cell stacks with little space or weight constraints. The configuration usually depends upon system design – whether heating is required in planar layers or curved surfaces – and calls for careful attention to thermal uniformity, electric insulation, as well as physical robustness. Advanced designs incorporate multi-zone heating with separate control loops for targeted pre-heating, supported by real-time temperature sensing and feedback methods. Insulating layers like mica, polyimide (Kapton), or silicone rubber are utilized to avoid heat loss and also protect surrounding components. The construction entails lamination, screen printing, or deposition of conductive tracks onto elastic substrates, followed by encapsulating the substrates for chemical and thermal protection. Power supply considerations are essential – designs need to fit available onboard auxiliary power while reducing electricity draw to protect fuel cell efficiency. Furthermore, thermal inertia and warm-up period are optimized through correct material thickness, heater geometry, and contact resistance minimization. Heaters might also possess rapid-response capabilities, including lower thermal mass and pulse width modification (PWM) controls to change power delivery in chilly climates. Ultimately, the design of resistance heaters for fuel cell systems must balance quick heating rates, longevity, along energy efficiency while adhering to safety and compactness requirements specific to stationary and automotive hydrogen applications.

4.5 APPLICATIONS OF ELECTRIC RESISTANCE HEATING

Electrical resistance heating finds vital uses in hydrogen fuel cell systems, particularly for addressing cold start challenges in sub-zero environments. Its main function is to pre-heat the fuel cell stack containing MEAs, GDLs, and flow area plates, to avoid ice formation and enable effective proton conductivity. Resistance heaters are commonly installed directly in the stack or adjacent thermal management systems within PEMFCs to provide a quick and uniform temperature increase in extremely sensitive lower temperature applications. They are utilized to increase inside stack temperatures from ambient or sub-zero temperatures on the operating range of 60–80°C, considerably minimizing the start-up period. Conformability and effective thermal output are the main reasons for utilizing thin-film and flexible heaters. Resistance heating is also utilized in pre-warming balance-of-plant (BoP) components like humidifiers, piping, and valves, which may freeze and hinder the hydrogen or air supply. For automobile fuel cell applications, resistance heating supports cabin preconditioning and battery thermal management, enhancing total vehicle readiness. Stationary power systems utilize resistance heaters to keep the idle-state warmth throughout intermittent operation. Electrical resistance heating within aerospace, as well as unmanned aerial vehicle (UAV) systems, guarantees dependable functioning in high-altitude and low-temperature environments. Technology is additionally employed in the lab testing as well as start-up characterization of the fuel cell stacks.

Considering the growing need for fuel cell installation in cold regions, resistance heaters are part of the standard design in commercial PEMFC systems. Control algorithms generally govern their integration by maximizing energy usage and reducing thermal gradients. The technology's flexibility and comparatively low cost make it an indispensable part of contemporary hydrogen fuel cell cold start approaches.

4.6 ENERGY EFFICIENCY AND OPTIMIZATION

Energy efficiency and optimization are essential considerations in the use of electric resistance heating for cold start solutions in hydrogen fuel cell systems. Fuel cells in sub-zero operations are subject to issues like ice formation and increased internal resistance and need fast and efficient thermal management. Even though simple and effective, electrical resistance heating may be energy-intensive if not managed correctly. To ensure minimal parasitic power losses, optimization methods have to concentrate on improving heater design, controlling heat distribution, and integrating intelligent management methods. Advanced materials with good thermal conductivity and low electric resistance, like nanoscopic films and carbon-based composites, can substantially increase the efficiency of resistance heating. Additionally, zonal heating, whereby only crucial parts like the MEA or flow fields are targeted, reduces total energy usage. Thermal insulation components around the fuel cell stack reduce heat loss, preserve the heat created, and lower the power input needed. Adaptive thermal management systems employing real-time temperature sensors and feedback loops can dynamically modify the heater output depending on external factors and system demands. Furthermore, pre-heating the stack using residual waste heat from prior activities or even keeping heat in PCMs can complement resistance heating and lower power demands. Integrating predictive algorithms that anticipate environmental conditions and operational schedules allows for preemptive thermal management, further optimizing energy usage. Battery management systems (BMSs) should also be harmonized with the heater load to prevent too much drain, ensuring sustained system performance. Furthermore, strategies like pulse heating or intermittent operation can optimize power usage without compromising heating success. Research is examining hybrid approaches that blend resistance heating with various other cold start methods, like catalytic or induction heating, to balance performance and energy efficiency. The ultimate objective is to guarantee dependable cold starts while incurring the very least power expense possible, making sure the auxiliary heating unit enhances instead of detracts from the overall energy efficiency of the hydrogen fuel cell vehicle. The improvement of energy consumption in cold start systems will probably be further improved by future developments in intelligent materials and AI-enabling thermal control.

4.7 CONTROL AND MONITORING SYSTEMS

Electrical resistance heating strategies in hydrogen fuel cell cold starting solutions require effective control and monitoring methods. These systems make certain that the heating elements work within optimum parameters to pre-heat the fuel cell stack efficiently and safely under sub-zero conditions. Thermal sensors, including thermocouples and resistance temperature detectors (RTDs), are the heart of these systems and constantly measure the temperature of different parts, for example, the MEA, the bipolar plates, and humidifiers. Feedback coming from these sensors enables dynamic modification of power input to the heaters via closed-loop control methods, preventing overheating and ensuring consistent thermal distribution. Microcontrollers or embedded systems operating thermal management algorithms are usually found in contemporary control units, enabling adaptive and predictive heating adapted to ambient temperature, start-up history, fuel cell load conditions, and so forth. Integration with BMSs enables optimum power distribution between traction needs and heating demands. Smart monitoring systems also offer real-time diagnostics, fault logging, and information logging to enable predictive maintenance and enhance system reliability. CAN (Controller Area Network) protocols or some other vehicle communication interfaces are usually employed for seamless integration into the vehicle's electronic control architecture. Redundant temperature sensors and hardware-level fail-safes, including thermal cutoffs or fuses, also improve safety. Advanced models utilize machine learning methods to anticipate the thermal behavior of fuel cells during cold start, further optimizing energy usage and start-up time. For operator visibility, real-time temperature profiles and heater status may be shown through graphic user interfaces (GUIs) or vehicle dashboards. Remote monitoring and over-the-air (OTA) update features are becoming increasingly prevalent, particularly in fleet activities or attached vehicle platforms. Generally, control and monitoring systems are indispensable for ensuring that electric resistance heating contributes successfully to safe, dependable, and energy-efficient cold starts in hydrogen fuel cell systems.

4.8 SAFETY AND RELIABILITY OF RESISTANCE HEATING SYSTEMS

Electrical resistance heating solutions for hydrogen fuel cell cold start applications must be reliable and safe. The integration of heating components near fuel lines or inside the fuel cell stack is governed by strict thermal and electric safeguards because hydrogen is a highly flammable as well as explosive substance. Resistance heaters have to be designed to operate inside tightly controlled temperature thresholds to stay away from overheating, which could degrade membrane materials or trigger unintended reactions. The system should include thermal insulating material and shielding to avoid heat damage and stay away from exposure to the surrounding parts. Reliability is attained by utilizing materials like nichrome, Kanthal, or advanced carbon-based composites, which can withstand repeated thermal cycling without degradation. Continuous monitoring as well as response requires safety circuits including temperature sensors, current limiters, and feedback control loops. Runaway heating scenarios are frequently prevented by redundant control systems and safety shut-offs. In automobile applications, compliance with industry standards like ISO 26262 for purposeful safety and IEC 61508 for electric systems is crucial. The validation includes insulation resistance tests, fault detection diagnostics, and thermal runaway simulations. To prevent interference, the heating system should be electromagnetically in tune with the other electric parts of the automobile. The additional battery of the fuel cell, which powers the heater, also has to be protected from short circuits or voltage spikes. Accelerated life tests and exposure to intense environmental conditions, including sub-zero temperatures and humidity cycling, validate long-term reliability. The design incorporates safety margins to account for component aging and thermal fatigue. In case of any problem, fail-safe modes enable the system to turn off or enter "low-power" mode. Maintenance-free operation and little performance drift over time are attractive features. By guaranteeing both intrinsic and operational safety, resistance heating systems can offer reliable functionality across the fuel cell's service life. Ultimately, integrating reliable and safe heating solutions enhances total system robustness, driver safety, and user self-confidence, particularly in cold climate operations.

4.9 CURRENT STATE OF IMPLEMENTATION AND PROGRESS IN ELECTRIC RESISTANCE HEATING

One of the most mature and well-known solutions to cold start issues in hydrogen fuel cell systems (especially in PEMFCs) is electric resistance heating. Its use is necessitated by the fact that it is simple, reliable, and can quickly produce energy through the Joule effect, thus it is practical during the process of melting ice formation and also in a start-up at low temperatures. Toyota Mirai and Honda Clarity are examples of commercial fuel cell cars that have already incorporated electric resistance heating components or subsystems, including a humidification component and coolant loop components into the stack, or stack components. The newest development is more about super-efficient heating secured with minimum energy usage to guarantee car range and durability. Due to their small form factor and fast thermal response, thin-film and foil-based microheaters are being used as localized heating, especially at the MEA or GDLs, to speed up electrochemical reactions. Developments in materials science have allowed the creation of high-resistivity alloys and composites based on carbon with enhanced thermal qualities and durability. It may be coupled with smart control systems and sensors so that power can be dynamically modulated using real-time feedback of the temperature changes, enhancing both efficiency and safety. Automotive OEMs and fuel cell stack makers are investigating the applications of multi-zone resistance heating together with thermal insulation to solve the problem of equal temperature distribution and the thermal gradient reduction. Work is also underway to address the power demand at start-up by putting resistance heaters and auxiliary batteries or supercapacitors together and thus pre-heating effectively even at very low temperatures. Resistance-heated stacks have been proven in studies conducted in harsh winter conditions of countries like Canada, Japan, and other parts of the European continent. Initiatives are going on to optimize heater design for decreased mass, enhanced longevity, and compatibility with fast cycling conditions inherent to frequent starts and stops. Furthermore, organizations like SAE and ISO are trying to standardize testing protocols and productivity indicators for cold start warming systems. The future direction includes integrating resistance heating with hybrid thermal management strategies, like using PCMs or heat pumps, to reduce dependence on electric energy. As the hydrogen economy expands, the demand for energy-efficient and strong cold start solutions will probably grow, with electric resistance heating still playing a crucial role in system readiness and operational reliability in a range of climatic conditions.

4.10 CONCLUSIONS

The technology of electric resistance heating is essential in addressing the persistent cold start issues experienced by hydrogen fuel cells, particularly in portable and automotive PEMFC systems. Its capability to offer quick, localized, and energy-efficient heating guarantees system readiness in freezing environments, preventing membrane damage, ice formation, and reduced electrochemical performance. The integration of advanced heater types – including thin film, PTC, carbon-based, and flexible printed heaters – demonstrates how innovation in design and materials can significantly enhance thermal management without compromising fuel cell compactness or efficiency. Resistance, thermal conductivity, and the durability against hydrogen embrittlement were further improved by technological advances, especially in stainless steel alloys, Kanthal, and carbon nanostructures. Effective control and monitoring systems, together with real-time thermal feedback and safety mechanisms, guarantee precise temperature regulation and system protection. With growing commercial adoption and ongoing research centered on energy optimization, smart control, and hybrid heating strategies, electrical resistance heating is set to remain a primary component of future hydrogen fuel cell systems. The scalability, compatibility, and adaptability of its fuel cell architectures make it a crucial enabler of dependable cold starts, which will eventually help drive the wider shift toward hydrogen-powered clean energy solutions in most circumstances.

REFERENCES

Barnoon, P. (2021). Modeling of a high temperature heat exchanger to supply hydrogen required by fuel cells through reforming process. *Energy Reports*, 7, 5685–5699. https://doi.org/10.1016/j.egyr.2021.08.171

Cao, Q., Min, H., Sun, W., Zhao, H., Yu, Y., Zhang, Z., & Jiang, J. (2024). A method of combining active and passive strategies by genetic algorithm in multi-stage cold start of proton exchange membrane fuel cell. *Energy*, 288, 129794. https://doi.org/10.1016/j.energy.2023.129794

Chen, Y., Zhang, S., Chung-Yen Jung, J., & Zhang, J. (2023). Carbons as low-platinum catalyst supports and non-noble catalysts for polymer electrolyte fuel cells. *Progress in Energy and Combustion Science*, 98(June), 101101. https://doi.org/10.1016/j.pecs.2023.101101

Cui, B., Jin, J., Cheng, L., Xue, X., Liang, Y., & Wang, W. (2023). Enhanced response speed of SAW based hydrogen sensor employing a micro-heater. *International Journal of Hydrogen Energy*, 48(45), 17339–17348. https://doi.org/10.1016/j.ijhydene.2023.01.233

Duan, C., Kee, R. J., Zhu, H., Karakaya, C., Chen, Y., Ricote, S., Jarry, A., Crumlin, E. J., Hook, D., Braun, R., Sullivan, N. P., & O'Hayre, R. (2018). Highly durable, coking and sulfur tolerant, fuel-flexible protonic ceramic fuel cells. *Nature*, 557(7704), 217–222. https://doi.org/10.1038/s41586-018-0082-6

Duffin, T. G., Palomares, K. B., & Zinkle, S. J. (2023). Corrosion and microstructural characterization of molybdenum cermets following hydrogen exposure up to 2630K. *Journal of Nuclear Materials*, 586(August), 154692. https://doi.org/10.1016/j.jnucmat.2023.154692

Elkhatib, R., Kaoutari, T., & Louahlia, H. (2024). Green hydrogen energy source for a residential fuel cell micro-combined heat and power. *Applied Thermal Engineering*, 248(PA), 123194. https://doi.org/10.1016/j.applthermaleng.2024.123194

Gehmlich, N. T., Fuerst, T. F., Gietl, H., Taylor, C. N., Rittenhouse, J., Wen, H., & Cinbiz, M. N. (2025). Hydrogen permeation in iron-chromium-aluminum (FeCrAl) alloys and the effects of microstructure and surface oxide. *Journal of Nuclear Materials*, 603(June), 155397. https://doi.org/10.1016/j.jnucmat.2024.155397

Jiang, W., Song, K., Zheng, B., Xu, Y., & Fang, R. (2020). Study on fast cold start-up method of proton exchange membrane fuel cell based on electric heating technology. *Energies*, 13(17), 1–25. https://doi.org/10.3390/en13174456

Liu, Q., Tian, B., Luo, C., Liang, J., & Wu, W. (2020a). Printed flexible heaters-based thermotherapy platform for multiduty thermal management. *Advanced Materials Technologies*, 5(8), 1–9. https://doi.org/10.1002/admt.202000278

Liu, Z., Zhou, M., Chen, M., Cao, D., Shao, J., Liu, M., & Liu, J. (2020b). A high-performance intermediate-to-low temperature protonic ceramic fuel cell with in-situ exsolved nickel nanoparticles in the anode. *Ceramics International*, 46(12), 19952–19959. https://doi.org/10.1016/j.ceramint.2020.05.062

Liu, Q., Tian, B., Liang, J., & Wu, W. (2021). Recent advances in printed flexible heaters for portable and wearable thermal management. *Materials Horizons*, 8(6), 1634–1656. https://doi.org/10.1039/d0mh01950j

Nady, H., El-Rabiei, M. M., Samy, M., Deyab, M. A., & Abd El-Hafez, G. M. (2021). Novel Ni–Cr-based alloys as hydrogen fuel sources through alkaline water electrolytes. *International Journal of Hydrogen Energy*, 46(70), 34749–34766. https://doi.org/10.1016/j.ijhydene.2021.08.056

Osinkin, D. A. (2023). Identification of gas diffusion phenomena on highly active Ni–ceramic anodes using the DRT technique. *Journal of Power Sources*, 571(February), 233085. https://doi.org/10.1016/j.jpowsour.2023.233085

Park, J., Jang, I., & Kim, H. J. (2019). Crumpled carbon nanotube thin film heaters for high sensitivity hydrogen sensing. *Proceedings of IEEE Sensors*, 2019-October, 1–4. https://doi.org/10.1109/SENSORS43011.2019.8956523

Singh, R., Oberoi, A. S., & Singh, T. (2024). Effect of cylindrical heat pipes on electrochemical hydrogen adsorption within a carbon electrode employed in a modified reversible PEM fuel cell. *International Journal of Hydrogen Energy*, 52, 406–419. https://doi.org/10.1016/j.ijhydene.2023.09.045

Topcu, A., Aydın, K., & Çelik, S. (2022). The effect of the heater and the pump operation durations on the cold start performance of the coolant heating assisted PEM fuel cell application. *International Journal of Energy Research*, 46, 23939–23962. https://doi.org/10.1002/er.8692

Wang, Q., Luo, T., Tong, Y., Dai, M., Miao, X. Y., Ricote, S., Zhan, Z., & Chen, M. (2022). Large-area protonic ceramic cells for hydrogen purification. *Separation and Purification Technology*, 295(May), 121301. https://doi.org/10.1016/j.seppur.2022.121301

Wang, R., Ding, L., Pan, Y., Zhang, X., Yang, M., & Zhu, C. (2024). Ion implantation combined with heat treatment enables excellent conductivity and corrosion resistance of stainless steel bipolar plates for hydrogen fuel cells. *Materials*, 17(4). https://doi.org/10.3390/ma17040779

Xia, G., Habibullah, Xie, Q., Huang, Q., Ye, M., Gong, B., Du, R., Wang, Y., Yan, Y., Chen, Y., & Wu, C. (2024). Recent progress in carbonaceous materials for the hydrogen cycle: Electrolytic water splitting, hydrogen storage and fuel cells. *Chemical Engineering Journal*, 495(June), 153405. https://doi.org/10.1016/j.cej.2024.153405

Yang, Y. F., Hao, J. M., Zhang, W., Shen, Y. T., Zhou, R., Chen, H., & Cheng, W. L. (2024). High-precision adaptive temperature control performance of thin positive temperature coefficient materials with ultra-high resistance-temperature coefficient. *Applied Thermal Engineering*, 253(March), 123767. https://doi.org/10.1016/j.applthermaleng.2024.123767

Zhou, X., Zheng, X., Li, M., Fei, F., Cao, X., Dong, X., Li, L., Yuan, N., & Ding, J. (2024). Efficient power generation based-on carbon yarn in coaxial hydrogen peroxide fuel cells. *Materials Today Physics*, 44(January), 101432. https://doi.org/10.1016/j.mtphys.2024.101432

5 Self-Heating through Load Cycling

5.1 INTRODUCTION

Proton Exchange Membrane Fuel Cells (PEMFCs) are hydrogen fuel cells that call for considerable functional challenges at sub-zero temperatures because of the freezing of water in the membrane and electrodes, which may compromise start-up and efficiency. Self-heating through load cycling, a technique where the fuel cell creates heat internally by quickly switching between high as well as low electric loads, is a promising approach for this issue. Load cycling employs the internal resistances and electrochemical responses of the cell to create localized heat, while standard external heaters add weight and complexity. During cold starts, this technique is particularly advantageous as it enables the system to uniformly and gradually warm up crucial elements – such as the catalyst layers, gas diffusion layers (GDLs), and membrane – from inside, minimizing thermal stress and liquid buildup. Internal temperature increase is brought on by ohmic heating, electrode-activated losses, and hysteresis phenomena in materials resulting from the alternating current load throughout load cycling. The heat produced during cycling could be controlled to guarantee the safe and effective thawing of frozen components by carefully monitoring the cycling parameters – frequency, length, and amplitude. Integrating self-heating via load cycling in fuel cell start-up protocols not only improves cold start reliability but also enhances system compactness, reduces system auxiliary power needs, and extends the functional range of fuel cell vehicles even in cold weather. Heat generation throughout the load cycle in a hydrogen fuel cell can be attributable to many mechanisms.

5.1.1 Joule Heating

Joule heating, also called Ohmic heating, is a crucial mechanism for enabling cold start strategies inside hydrogen fuel cells, especially PEMFCs, by internally creating heat through electric resistance. The fuel cell elements – the polymer electrolyte (PE), electrodes, bipolar plates, as well as contact interfaces – encounter inner electrical resistance because of current flow, leading to ohmic heating. The equation governs the heat produced.

$$Q = I^2Rt$$

where

I = Current (A),
R = Total internal resistance (Ω), and
t = Time (s)

DOI: 10.1201/9781003657866-5

The proton-conducting membrane, in addition to the electron-conducting layers, creates heat which increases with increased currents, thus making this heat particularly helpful during load cycling. The membrane resistance is frequently modeled as a function of membrane area in PEMFCs.

$$R = \frac{L}{\sigma A}$$

where

L = Membrane thickness (m)
A = Active area (m^2)
σ = Proton conductivity (S/m)

Resistance and heat generation are further aided by a substantial decrease at sub-zero temperatures. Ohmic heating is utilized during cold starts at sub-zero temperatures to quickly raise the cellular temperature, therefore stopping ice buildup, which can easily block electrode pores and decrease performance. Newer strategies, including the Alternate Hydrogen Pump (AHP), use limitless ohmic heating without the generation of water, leading to fast, effective, as well as durable cold starts even after thousands of cycles, with no degradation found at close to −30°C. Numerical and experimental studies confirm that optimizing current loading and managing the first present slope are essential, because higher slopes raise ohmic resistance – particularly in the anode catalyst layer – impacting both heating rate as well as start-up success. Direct loading the stack using internal ohmic heating allows faster heating rates than outside techniques, but must be handled carefully to stay away from too much resistance and voltage drops. External pre-heating, like metal hydride reactors or phase change materials, can enhance cold start reliability and energy efficiency, lessening start-up time while minimizing electricity consumption. Literature study highlights that internal ohmic heating is among the best methods for fast cold starts, but optimal results are attained when it is incorporated with various other thermal management solutions. Overall, ohmic heating remains central to overcoming the cold start challenge in hydrogen fuel cells, supporting their broader adoption in cold climates.

5.1.2 Electrochemical Activation Losses

These electrochemical activation losses are generally brought on by ice formation in the catalyst layer and GDL during the cold start of hydrogen fuel cells, which stops reactants from reaching active sites and decreases the electrochemically active surface area. The ice coverage decreases cell voltage and performance considerably, because the reduced catalytic sites hinder the electrochemical reactions needed for power generation. The energy barrier for starting the electrochemistry should be overcome by the voltage called activation overpotential (η_{act}). The Butler–Volmer equation may be used to describe it, linking the current density to the activation excess potential.

$$j = j_o \left[\exp\left(\frac{\propto_a F \eta_{act}}{RT} \right) - \exp\left(-\frac{\propto_c F \eta_{act}}{RT} \right) \right]$$

where

j = Current density (A/m^2)
j_o = Exchange current density (A/m^2)
η_{act} = Activation overpotential (V)
$\propto_a, \propto_c$ = anodic and cathodic charge transfer coefficients
F = Faraday's constant
R = Universal gas constant
T = Absolute temperature (K)

The level of activation losses is impacted by catalyst layer thickness, ionomer-to-carbon ratio, and also the distribution of ice. Thicker catalyst layers withstand much more ice just before failure, while a greater ionomer content improves performance loss and decreases porosity. Localized degradation, including crack formation, catalyst particle development, and ionomer agglomeration, could be due to non-uniform ice formation downstream in the cell, which can increase activation losses and long-term longevity difficulties. To bring down losses, techniques consist of balancing start-up current and hydrogen flow arrangements and utilizing strategies like AHP to create internal heat without water creation, decreasing ice formation and degradation. Also, utilizing supercooled water rather than ice at start-up can help membrane hydration and stop blockage, enabling a successful cold start. Understanding and controlling the electrochemical activation losses is vital to raising cold start reliability and extending the working lifetime of hydrogen fuel cells in sub-zero conditions.

5.1.3 Double-Layer Charging and Discharging

The dual-layer charging and discharging mechanism in the electrode–electrolyte interface is essential in determining start-up results, in addition to productivity during the cold start of hydrogen fuel cells, particularly at sub-zero temperatures. The double layer may be regarded as a short-term charge reservoir that can impact the original voltage response and current distribution needed for managing water and ice formation in the cell. The solid electrode, along with the electrolyte, is joined by an electric double layer (EDL). Whenever a potential is applied across a fuel cell, ions gather around the electrode surface area, counterbalancing the electronic charge on the surface. A double-layer capacitance $\left(C_{dl}\right)$ could be utilized to model the EDL analogous to a capacitor.

$$C_{dl} = \frac{\varepsilon_r \varepsilon_o A}{d}$$

where

ε_r = Relative permittivity of the electrolyte
ε_o = Vacuum permittivity
A = Surface area of the electrode (m^2)
d = Effective thickness of the double layer (m)

The current and voltage throughout the cell differ dynamically throughout load cycling. The EDL then charges and discharges continuously due to this. Both numerical and experimental research suggest that modifying current loading methods and pre-heating methods, like coolant or thermochemical metal hydride pre-heaters, can bring down cold start time and so improve voltage stability by encouraging consistent two-fold layer charging and reducing ice blockages in the catalyst layer and the GDL. The control of the double layer throughout start-up helps regulate its speed of creation and phase transitions, lessening the possibility of ice formation, which can obstruct reactant flow and compromise performance. Additionally, techniques that maintain membrane hydration and increase even heat distribution further improve double-layer dynamics, resulting in faster and reliable cold starts even at −30°C. By analyzing the interactions between electrochemical two-fold layer behavior, thermal control, and water management, these solutions enhance the longevity and business viability of hydrogen fuel cells in cold climates.

5.1.4 Hysteresis and Microstructural Friction

Hysteresis and microstructural friction are essential self-heating systems in hydrogen fuel cell cold starts. This phenomenon originates largely from cyclic electrochemical, thermal, and physical loading during the start-up stage, particularly when load cycling is used to induce internal heating. The fuel cell is subjected to reconfigure, contraction, and periodic expansion of structural parts, especially of the PE, GDLs, as well catalyst layers, during cycles. Microstructural modifications are not entirely reversible and lead to hysteresis gains that dissipate energy as heat. In hydrogen fuel cells, especially proton exchange membrane (PEM) types, ice formation and microstructural modifications trigger hysteresis and friction, along with other consequences, at the microscopic level, which impede performance during cold start. Microstructural harm in the cell's porous layers as well as the membrane, which includes loss of hydrophobicity and alterations to gas diffusion pathways, can be irreversible and eventually result in increased resistance and performance decay over several freeze–thaw cycles, which are closely associated with microstructural friction and hysteresis in cellular behavior. Self-heating throughout load cycling is aided by hysteresis and microstructural frictional mechanisms, which help melt frozen water and return the fuel cell to its working temperature. The mechanism includes three stages.

5.1.4.1 Step 1: Electrochemical Hysteresis

Self-heating depends upon electrochemical hysteresis, which transforms energy losses into heat through cyclical loading. The cell voltage does not follow a regular symmetric path since the current is varied quickly throughout load cycling because of delayed electrochemical responses such as slow activation kinetics, water transportation limits, and membrane hydration lag, particularly at sub-zero temperatures. A voltage–current ($V - I$) hysteresis loop is created by this action, where power is lost as heat within the enclosed area. Asymmetric behavior is also brought on by charging and discharging of the electrode chemical double layer and by the accumulation or redistribution or redistribution of water. The equation may be used to compute the heat created each cycle mathematically.

$$Q_{\text{hyst}} = \int V(I)\,dI$$

The internal heat resides near the catalyst and membrane regions, enabling frozen water to thaw, restore proton conductivity, and efficiently warm the cell without the intervention of external heating elements. Electrochemical hysteresis could be utilized to obtain dependable and energy-efficient cold starts for PEM fuel cells when integrated into controlled load cycling.

5.1.4.2 Step 2: Mechanical Hysteresis of the Membrane

The membrane in a PEMFC goes through physical hysteresis because of its viscoelastic behavior under diverse hydration and temperature conditions, leading to cyclic mechanical energy dissipation. The significance of this takes place during cold start, when cyclic swelling and contraction of polymer electrolyte membranes create inner heating to melt ice and propel start-up. The ensuing stress–strain action produces a hysteresis loop. The location inside the loop gives the energy lost as heat each cycle.

$$Q_{\text{mech}} = \int \sigma(\varepsilon)\,d\varepsilon$$

where

σ = Stress in the membrane (Pa)
ε = Strain in the membrane
Q_{mech} = Heat energy dissipated due to mechanical hysteresis (Pa)

Mechanical hysteresis also has a thermal impact on PEM fuel cells at cold start, as even small quantities of internally produced heat help to thaw frozen components and regain membrane conductivity. Localized heat is expelled while the membrane goes through cyclical swelling and contraction under hydration modifications and load cycling, releasing energy lost during viscoelastic deformation. The membrane electrode assembly (MEA) is rehydrated by this heat, which warms the MEA, melts ice in vital locations, and helps in restoring proton conductivity. The degree of mechanical hysteresis and the heat creation it leads to are impacted by several factors. Increased hydration levels lead to higher swelling and strain, raising the hysteresis loop region and therefore more heat. Temperature also plays a role, since viscoelastic behavior intensifies with rising temperatures up to the membrane's glass transition point. The hysteresis occurrence is amplified by thicker membranes, more rigid mechanical restrictions inside the fuel cell's stack, and quicker cycling frequencies. This phenomenon is a dispersed passive source of heat that complements other heating mechanisms like ohmic and electrochemical loss under real-world cold start conditions. The presence of membrane deformation cycles and the localized heating effects in dynamic operation were confirmed via experimental studies, including strain gauges and infrared thermography, as well as electrochemical impedance spectroscopy. The significance of utilizing mechanical hysteresis has been increasingly recognized by automotive fuel cell developers as a crucial part of an integrated self-heat approach for stable sub-zero start-up without external heat sources.

5.1.4.3 Step 3: Microstructural Friction at Interfaces

The MEA may be the laminated framework that holds the core parts in PEMFCs, which comprises the membrane, catalyst layers, and GDLs. Micromovements and mechanical interactions happen at the interfaces between these layers throughout cold start, particularly during load cycling. Microstructural friction is the process by which physical energy is transformed into heat because of inner shearing and sliding at layer interfaces due to microscopic moves. This subtle friction creates localized heat, which helps self-heating by helping the thawing of icy materials and warming the cell for start-up. Water within the MEA of a PEM fuel cell could freeze in many locations during sub-zero operation, including catalyst layer pores, membrane surfaces, and even microcracks or vacancies. The freezing and subsequent thawing of this water causes different expansion as well as contraction of adjacent layers because of the mismatched thermal expansion coefficients of the constituent substances. During load cycling, which entails switching between high and low current densities occasionally, cyclic electrochemical pressure and pulsations take place together. Dynamic effects bring about compression and relaxation of porous layers, expansion as well as contraction of the membrane, and relative slippage at the interface. These physical actions combined create microscopic shear forces at interfaces that create interfacial friction. This particular mechanism dissipates energy as heat and contributes to the self-heating needed for excellent cold starts. Frictional heat generated at an interface is analyzed at the microscale and is based on friction.

$$Q_{\text{friction}} = \mu . F_N . \Delta x$$

where

Q_{friction} = Heat generated due to friction (J)
μ = Coefficient of friction between layers
F_N = Normal force pressing the layers together (N)
Δx = Relative sliding displacement per cycle (m)

Localized self-heating is brought on by interfacial friction among important MEA layers, such as the membrane–catalyst interface, catalyst GDL, and GDL flow area plate during hydrogen fuel cell cold starts. Repeated mechanical interactions occur at these interfaces because of thermal excitation mismatches, ice buildup, and gas pressure variations caused by load cycling. Despite the little heat output per event, the cumulative impact throughout the whole cell area and throughout several cycles is essential for thawing ice, replenishing the membrane, and restoring the gas routes needed for cold start results. The interfacial heat raises temperature uniformity inside the cell and supplants main sources, including ohmic and activation losses. The intensity of frictional heating depends upon factors like material roughness, elastic modulus discrepancies, hydration or ice status, compression pressure, and load cycling profile. This effect is investigated and quantified using experimental tools like infrared thermographic analysis, dynamic mechanical analysis, and finite element modeling. Interfacial friction could be enhanced in design by utilizing textured or compliant materials and specialized cycling protocols to enhance inner heating without mechanical harm. Overall, interfacial friction provides an invaluable, passive mechanism to support fast and dependable cold starts in PEM fuel cells while reducing the need for external heating systems.

5.1.5 Water Phase Change and Evaporation

The phase change and evaporation of water during the cold start of hydrogen fuel cells, particularly PEMFCs, are essential to start-up results and productivity. Water created while in operation may be in a supercooled liquid or icy state at below-zero temperatures. Ice could block oxygen pathways and cause a start-up malfunction if it forms as well as builds up, particularly in the cathode catalytic level. Supercooled water may thus disperse away, decreasing clog danger and enabling effective cold starts. Water builds up and freezes stepwise, often beginning at the cellular center, leading to shutdown as freezing progresses, as demonstrated in visualization studies. The consequent phase changes from ice to liquid and then vapor are essential to restoring cellular function. Phase changes assist internal heating by latent heat exchange, control temperature via thermal energy transfer throughout the cell, hydrate the membrane for effective ionic conduction, and remove excess water through evaporation, preventing flooding and ensuring stable function. Multiphase and mechanistic models have discovered that water transportation and distribution, as well as removal, affected by hydrogen convection and start-up methods, are essential for managing membrane water content and stopping icing formation. Mixtures of alternate hydrogen pumping and targeted purging (including secondary purge), along with accurate control of start-up current and voltage, can decrease membrane water content, stop ice formation, and also extend the functional window for effective cold starts. Further strengthening cold start dependability are thermal management systems that utilize phase change substances to keep stack temperatures above freezing. The phase change entails thermodynamics.

5.1.5.1 Melting of Ice (Solid to Liquid Transition)

The latent heat of fusion required to convert ice into water, which is calculated from

$$Q_{\text{melt}} = m.L_f$$

where

Q_{melt} = Heat required to melt ice (J)
m = Mass of ice (kg)
L_f = Latent heat of fusion for water (J/kg)

This heat must come from internal generation (e.g., ohmic heating, load cycling) to enable thawing.

5.1.5.2 Evaporation of Liquid Water (Liquid to Vapor Transition)

The latent heat of vaporization governs water evaporation, which is calculated from

$$Q_{\text{evap}} = m.L_v$$

where

Q_{evap} = Heat required to evaporate water (J)
m = Mass of ice (kg)
L_v = Latent heat of vaporization for water (J/kg)

Evaporation is helpful during fuel cell cold start operations because it rebalances thermal energy from cool, localized heat sources – and also enables water management by eliminating unnecessary liquid water from GDLs and flow areas.

5.1.5.3 Vapor Pressure and Evaporation Rate

The speed of evaporation is determined by heat and the vapor pressure gradient. The Hertz–Knudsen Equation is utilized in computing the mass flux of evaporated water.

$$\dot{m}_{\text{evap}} = \propto \sqrt{\left(\frac{M}{2\pi RT}\right)}\left(P_{\text{sat}}(T) - P_v\right)$$

where

m_{evap} = Mass flux of evaporated water (kg/m²·s)
$\propto$ = Accommodation coefficient (~0.05–1)
M = Molar mass of water (0.018 kg/mol)
R = Universal gas constant (8.314 J/mol·K)
T = Absolute temperature (K)
$P_{\text{sat}}(T)$ = Saturation vapor pressure at temperature T (kg/m-s^2)
P_v = Partial pressure of vapor in gas stream (kg/m-s^2)

5.1.5.4 Ice Formation and Melting Locations

Water may freeze or even melt in several areas of a fuel cell, like the catalyst layer (CL), where it is created, allowing it to freeze within pores; capillary forces keep water and ice in the GDL; the membrane retains water for proton conduction, and it is subject to ice damage; and also the flow channels, where bulk water frequently gathers and freezes. The melting rate of ice is associated with the local heat exchange and thermal conductivities of materials.

$$q = \frac{k}{d}\left(T_{\text{source}} - T_{\text{ice}}\right)$$

where

q = Heat Flux (W/m^2)
k = Thermal conductivity of the medium (W/(m.K))
d = Distance from heat source (m)
$T_{\text{source}}, T_{\text{ice}}$ = Source and ice temperatures (K)

Restoring membrane conductivity calls for water to stay liquid throughout a cold start, since frozen or dry membranes stop proton transport. Furthermore, stopping ice formation stops physical damage from expansion, which might crack catalyst layers or delaminate interfaces. This requires unobstructed gas flow, as ice in the flow channels is able to obstruct the delivery of oxygen and hydrogen. Moreover, effectively utilizing latent heat during freezing and melting optimizes internal heat management, enhancing overall cold start effectiveness.

The total heat requirement for a cold start includes the following:

$$Q_{\text{total}} = m_{\text{ice}}\, c_{\text{ice}} \left(0 - T_{\text{start}}\right) + m_{\text{ice}}\, L_f + m_{\text{water}} . c_{\text{water}} \cdot \left(T_{\text{target}} - 0\right)$$

where

c_{ice} = Specific heat of ice (~2.1 kJ/kg·K)
c_{water} = Specific heat of water (~4.2 kJ/kg·K)
T_{start} = Initial temperature (°C)
T_{target} = Target operating temperature (°C)

In PEM fuel cells, cold start presents numerous difficulties, including the limitation of melting speed in thicker ice layers or thermal conductivity zones, uncontrolled and unbalanced evaporation that could overdry the membrane, non-uniform heating that could bring about partial thawing and refreezing cycles, and the need for balancing water removal and membrane hydration. Effective control and design methods are necessary to address these problems. Load cycling and current pulse methods are utilized to control internal heating, while thermal and moisture sensors monitor water state; hydrophilic/hydrophobic GDLs are made to regulate water migration during phase transition, and multiphysics models are used to effectively mimic phase shift dynamics, thermal transfer, and fluid flow during cold starts.

5.2 MATERIAL BEHAVIOR UNDER LOAD CYCLING

For instance, the tenacity and performance of hydrogen fuel cell cold start solutions depend upon material behavior under load cycling throughout self-heating. Ice formation and uneven water distribution during cold starts could result in mechanical stress and degradation in the membrane layer as well as the catalyst layer, but advanced fuel cell designs and strategies like constant power start-up and coolant pre-heating, along with adaptive temperature regulators, relieve these issues by promoting consistent temperature and water distribution throughout the stack. As demonstrated by electrochemical impedance spectra and microscopic analysis, load cycling and start–stop procedures, especially with diluted hydrogen feeds, could accelerate degradation by increasing ohmic and mass transport resistance, leading to a need for careful operational control. Regardless of these difficulties, recent prototype fuel cell stacks have demonstrated only minor voltage drops and limited physical damage after accelerated cold start cycling, meeting automotive durability targets. Cold start strategies, including pre-heating and real-time adaptive control, can guarantee a successful start even at temperatures down to −40°C, but voltage inconsistencies because of incoherent heating, as well as water circulation, can happen with higher output energy during a cold start. The mechanical, electrochemical, and thermal behaviors of the materials involved are impacted by these conditions.

5.2.1 Membrane Behavior

The membrane in PEMFC undergoes significant stress during load cycling during self-heating within hydrogen fuel cell cold start scenarios, resulting in performance and durability compromises. The degradation of the membrane as well as catalyst layers is sped up by load cycling, which raises ohmic resistance, decreases the electrochemically active surface area, and lowers total cellular efficiency. In a cold start, particularly under continuous power or current, the water created in the cell might freeze near the membrane, causing ice formation, which disturbs conductivity and will physically harm the membrane because of volume changes. The first water content and membrane thickness are crucial factors influencing cold start success and subsequent membrane behavior, because they impact both ice formation and the redistribution of water within the cell. Self-heating techniques like the thermochemical metal hydride ozone generator or a thermochemical metal hydride pre-heater can minimize ice formation and also enhance voltage stability, therefore minimizing chemical and mechanical stresses on the membrane at cold start. Repeated freeze–thaw and load cycle events still lead to microstructural modifications and loss of hydrogen ions in the GDL, causing performance decay. A detailed experimental examination of the degradation mechanisms of hydrogen–oxygen PEMFCs under present cyclic loading was performed by Meng et al., providing crucial insights into performance loss, catalyst aging, and membrane damage. They conducted a 2500-cycle durability test using a 50 cm^2 PEM FC and discovered a 31.0% fall in voltage at 1200 mA/cm^2 and a 61.6% reduction of electrochemical surface area (ECSA). Characterization techniques including polarization curves, cyclic voltammetry (CV), EIS, SEM, EDS, and TEM demonstrated widespread platinum (Pt) particle aggregation, migration, and dissolution, particularly in the outlet regions caused by gas star The PEM underwent substantial chemical degradation, as exhibited by elemental loss (F, C, and S) as well as mechanical thinning brought on by repeated hydration–dehydration and free radical attacks. These results are particularly pertinent to hydrogen fuel cell cold start solutions where quick load transitions and low-temperature conditions resemble the stresses brought on by cyclic current loading. This investigation suggests that unstable gas flow, temperature gradients, and water imbalance are typical cold starts that accelerate catalyst degradation and membrane failure. The results emphasize the need for cold start methods that facilitate load application smoothly and good water and gas management, in addition to the creation of durable catalyst and membrane materials that can withstand thermal and electrochemical cycling. Moreover, the determination of localized degradation at outlet zones offers design insights for enhanced structural and thermal management, ultimately enhancing the dependability and lifetime of PEMFC systems in cold environments (Meng et al., 2022).

Figure 5.1 describes the electrochemical processes inside a PEMFC, showcasing the behavior of platinum (Pt) catalysts and ion migration, which directly influence membrane performance and degradation. At Pt catalyst locations on the anode side, hydrogen is oxidized to create protons (H^+) and electrons (some Pt and e^-) that dissolve into the ionomer as Pt^{2+} ions, especially during transitory conditions such as cold start or load cycling. These Pt^{2+} ions switch throughout the PEM, wherever they could go down to metal Pt by inbound electrons, resulting in Pt precipitation within the membrane. Platinum crossover harms the membrane's structural integrity and creates

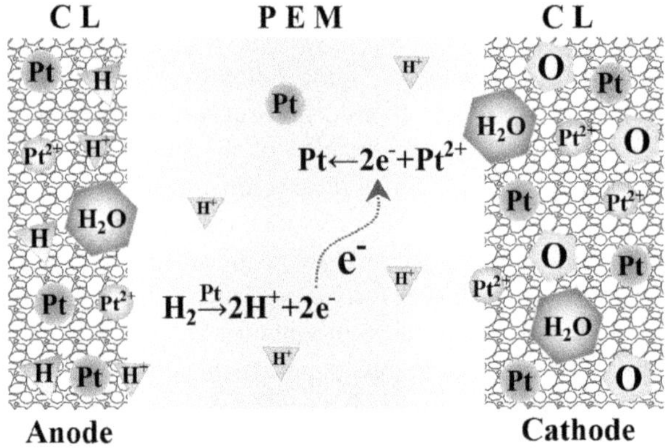

FIGURE 5.1 The dissolution, migration, and precipitation of Pt particles (Meng et al., 2022).

electrically conductive locations, which foster hydrogen crossover and decrease ion selectivity. Oxygen creates water on the cathode side when it comes together with electrons and protons, finishing the oxygen reduction reaction, but Pt^{2+} ions can reappear on the anode and form Pt bands, which trigger the formation of reactive oxygen species (ROS) like hydroxyl and hydroperoxyl radical groups. The membrane is chemically assaulted by these radicals, leading to thinning, cracking, and ultimately mechanical failure – effects that are exacerbated under cold start conditions because of water freezing, hydration–dehydration cycling, as well as with abrupt temperature shifts. Additionally, insufficient water management during cold starts can lead to ice buildup and swelling–shrinking stress, which intensifies membrane degradation.

5.2.2 Catalyst Layer (CL) Behavior

The CL also faces issues during the cold start of hydrogen fuel cells, particularly under load cycling and self-heat conditions, because ice formation and water management present important risks. Optimizing the CL pore structure and material distribution, like the use of electrospun nanofiber layers or ionomer loadings graded to platinum or ionomer, could enhance cold start ability by decreasing mass transport losses, enabling consistent ice deposition, and enhancing longevity against freeze–thaw injury. Self-heating loads produce internal heat, which can melt ice and ease start-up, while simultaneously exposing them to non-uniform ice development and localized degradation downstream in the CL, where critical success factors are CL thickness, ionomeric-to-carbon ratio, and pore size, all of which impact ice blockage and reactant flow. Important are self-heating strategies that balance current densities and heat production, because higher currents can accelerate warming while simultaneously increasing the possibility for ice-induced failure when poorly managed. Crack development and platinum particle development, along with ionomer aggregation, are degradation mechanisms during repeated cold starts, especially in areas with higher ice accumulation. Cold start performance could be enhanced through advanced modeling and real-time control of temperatures and load to avoid structural damage. Liu et al. evaluated the cold start performance of electrospun catalyst layers

(E-spun CLs) in PEMFC and discovered they outshine standard electro-sprayed catalyst layers (E-sprayed CLs). The E-spun CLs had a nanofiber structure with good porousness and substantial micropore distribution, leading to lower mass transport losses, better water management, and enhanced ECSA stability. The increased performance was linked to these features in sub-zero cold start situations, when supercooled product water freezes and ultimately causes structural damage and fuel cell failure. E-spun CLs, having a microporous structure, experienced a substantial reduction in freezing probability and absorbed and drained far more water, leading to greater survival times at 10°C and −20°C. Usually, E-spun CLs maintained structural integrity and also displayed only 8.5% ECSA loss following repeated cold starts, compared to 31.4% for E-sprayed CLs, showcasing their mechanical durability and resistance to platinum degradation. The freeze-resilient character of the E-spun CLs was confirmed by the lack of substantial catalyst aggregations or collapse, indicating a promising cold start solution for automotive applications running in cold weather (Liu et al., 2021).

5.2.3 Gas Diffusion Layer (GDL) Behavior

The GDL is crucial for water and heat management in hydrogen fuel cells and influences self-heating and cellular efficiency. Studies suggest that GDLs with bigger-sized pores and smooth pore size gradients enhance cold start performance because of decreased mass-transfer resistance and improved water and ice storage, which prevent cell degradation during self-heat phases. The quality of hydrophobicity is equally crucial – a moderate hydrophobicity aids water transfer at about −5°C, while much stronger hydrophobicity is beneficial at lower temperatures because it impacts water as well as ice action and thus prolongs stable operation. Pore-scale modeling reveals that when the cell self-heats, heat is transferred into the GDL, and improved porosity or even broader pores facilitate much better fluid water management, accelerating the removal of melted ice and supporting electrochemical reactions. Patterned wettability and dual-layer GDL design additionally enhance cold start by encouraging fast water movement and exclusion, lowering the risk of ice blockage. Hydrophobic coatings, along with microporosity, could influence water distribution and freezing behavior, therefore lowering variability in cold start outcomes and cell failure probability. The research by Wang et al. involved including patterned wettability into the MPL as well as the GDL of the MEA to boost cold start performance of PEMFCs by combining hydrophilic and hydrophobic zone designs of MPLs and GDLs to direct water flow, enhance drainage, and minimize ice formation during sub-zero start-up conditions. The authors utilized a bespoke experimental setup that imitated an internal fuel cell stack environment, demonstrating that while traditional configurations permitted effective start-up only at −10°C, their H-MPL/C-DDL and dual H-MPL models extended cold start capability to −12.5°C as well as −15°C, respectively. Even though the operation failed at −17.5°C, the hybrid components considerably postponed ice blocking as well as extended cellular functionality. This passive, material-based strategy offers an energy-efficient alternative to external heating, leveraging microstructural tuning of carbon-based layers to enhance water management, thermal distribution, and overall membrane resilience (Wang et al., 2021).

5.2.4 Bipolar Plate Behavior

The behavior of bipolar plates is essential to the self-heating performance of hydrogen fuel cells at cold starts and below-zero temperatures. Graphite or metallic bipolar plates possess significant thermal mass and will impact heat distribution, facilitating or hindering fast temperature rise in self-heating strategies. The AHP and integrated catalytic heating units have been created to enhance heat creation directly at the bipolar plates so that cool starts from −30°C to 40% can be attained with success, allowing heat to be effectively transferred to the MEA and preventing ice formation, which might obstruct active sites and affect performance. Bipolar plate design, as well as material characteristics like thickness, heat capacity, and surface coatings, plays a major role in heat dissipation rates, with slimmer plates in addition to materials with reduced heat capacity boosting cold start responsiveness. Vanadium oxide (V_2O_3) thin films, along with other advanced coatings, can offer extra self-heating by Joule heating, leading to decreased thaw times and power usage. The bipolar plate surface must have uniform heat distribution to prevent local dehydration or thermal degradation, therefore ensuring stable operation and extending fuel cell life. The self-heating solution developed by Jung et al. involves coating 316L stainless steel bipolar plates with V_2O_3 thin films, which display negative temperature coefficient (NTC) behavior to boost the cold start ability of PEMFCs. These films, created by a sol-gel dip-coating technique, showed reducing electric resistance with dropping temperatures, allowing effective Joule heating from −20°C to 80°C. The films had a resistance of ~0.28 Ω at −20°C, creating sufficient localized heat to melt internal ice at start-up, offering an energy-efficient and light alternative to traditional external heating systems. Using NTC-based self-heat allows for advanced thermal regulation to minimize power usage while staying away from membrane and catalyst damage due to freeze–thaw stress. By incorporating films directly onto bipolar plates, the bulky heating components are not needed, and the system becomes a lot more lightweight and reliable. Smart material coatings can allow energy-optimized and durable operation of PEMFCs in chilly environments, which is particularly crucial for automotive fuel cells, where fast, low-power cold starting functionality is crucial (Jung et al., 2011).

Wen et al. created a novel AHP technique for cold starting of graphite bipolar plates (GBPs) equipped with PEMFCs from temperatures as low as 30°C. In a pure hydrogen environment, this method utilizes alternating voltage to create internal heating through ohmic polarization, therefore stopping water development and ice accumulation, which usually slows cold beginnings. AHP offers a safer, lower-energy, and more granular thermal strategy compared to traditional external heating or hydrogen–oxygen combustion strategies. The study demonstrated that start-up from −30°C to 0°C was achievable in 7.2 min with improved performance linked to increased original membrane hydration, optimized switching frequency (>1 Hz), and an intermittent hydrogen supply technique to keep the membrane water content balanced with fuel availability. The AHP method is particularly essential for GBP-based PEMFS due to their higher thermal inertia when compared with metal-based systems, because it guarantees even heat distribution and powerful activation while staying away from freezing damage. The DOE cold start technology meets DOE cold start energy objectives and offers a practical, system-ready option for automotive and heavy-duty hydrogen fuel cells, which is prepared for integration (Wen et al., 2019).

5.3 APPLICATIONS OF SELF-HEATING THROUGH LOAD CYCLING

Self-heating via load cycling is a promising strategy for enabling rapid and efficient cold starts in hydrogen fuel cells, especially in automotive applications. The fuel cell creates internal heat when it cycles the power load intentionally, therefore raising its temperature and enhancing performance at low ambient temperatures. Using adaptive methods that exploit maximum current density at start-up has been found to accelerate membrane hydration and temperature rise, leading to quicker and more stable cold starts without substantial energy or hardware cost increases. Self-heating by load cycling can substantially shorten start-up time and energy use when compared with traditional techniques like external coolant pre-heating, particularly when coupled with different methods like phase shift materials or even optimized thermal management. Research indicates that combining self-heating with sophisticated control algorithms or even hybrid heating systems enhances cold start effectiveness, leading to reliable operation even at temperatures close to 40°C. The improvement of these new developments is essential for boosting the business viability of hydrogen fuel cells in cold weather, and they've already been employed in applications including fuel cell buses used for the Beijing Winter Olympic Games. Load cycling offers self-heating, and it is crucial for cold start functionality in most hydrogen fuel cell applications. Internal current pulsing within automotive fuel cell systems, like those in the Toyota Mirai and the Hyundai Nexo, quickly raises the stack temperature in chilly climates without external heaters, reducing start-up times and improving energy efficiency. Buses and trucks operating within sub-zero conditions also reap the benefits of this strategy, because it guarantees dependable early morning starts with no big heating units. In backup power and stationary fuel cells, particularly for telecom towers and distant facilities, load cycling allows autonomous thermal regulation after long periods of inactivity, preserving system functionality in the winter. This self-contained heating mechanism also helps portable fuel cell products used in military or emergency operations to offer dependable functionality in off-grid or freezing conditions. Load cycling within rail and marine applications guarantees pre-start thermal conditioning for safe and effective ignition because of the harsh temperatures encountered by hydrogen-powered systems. The same concept extends to aerospace and space applications, where extreme background temperatures make internal self-heating via load cycling important for starting power systems without outside thermal management. Finally, in lab test benches and diagnostics, load cycling is utilized to mimic cold start scenarios, helping evaluate membrane integrity, catalyst resilience, and, in general, stack durability under freeze–thaw cycles. These applications show the versatility and essential value of load cycling self-heating for a strong, scalable cold starting option for hydrogen fuel cells.

5.3.1 Automobile Fuel Cell Systems

The sandwich self-heating structure-based lithium-ion battery system (LIBS) was proposed as well as implemented by Cheng Lin et al. to enhance the cold start performance of HFCVs, particularly in extremely low temperatures like those experienced during the 2022 Beijing Winter Olympic Games. The concept behind the innovation revolves around a dual-circuit design that utilizes tiny, high-efficiency metal heating sheets embedded between battery cells to provide quick, uniform, and independent heating of battery packs from −30°C to operating conditions in under nine minutes. Internal heating performed better than standard Positive Temperature Coefficient (PTC) heating techniques with an 8.94 times faster temperature rise, greater heating efficiency (78.75%), and lower energy usage while having minimal effect on battery life and power density. This heating system assisted smaller-capacity batteries in supplying the power necessary to start as well as stabilize the fuel cell set (FCS), causing a drop in total system weight and cost. The sandwich self-heating design for cold start hydrogen fuel cell applications enables the battery to quickly provide power to the balance of plants (BOP), allow FCS ignition, and safely get output during startup, even at sub-zero degrees. The closed-loop heating system proves invaluable for automobiles in chilly weather because it guarantees a dependable cold start without outside energy, and it is scalable for both BEVs and HFCVs (Lin et al., 2023).

The sandwich's self-heating structure is discussed in the passage and shown in Figure 5.2. The advanced thermal management design (Figure 5.2) was produced to facilitate quick and efficient heating of LIBSs for applications like cool starts of hydrogen fuel cell vehicles (HFCVs). It's intended to be energy independent with no outside heat source. Rather, it employs internal discharge using embedded metal heating sheets placed on each battery cell's two largest surfaces. These sheets cover over 90% of the cell's surface area, guaranteeing a short maximum heat transfer distance (only half the cell's thickness), which allows incredibly fast and uniform warming. The battery pack is assembled with the cells and heating sheets stacked in an alternate way to produce a "sandwich" configuration. Throughout the heating process, two concurrent sources of heat are activated: The outside heating sheets transform electric power into Joule heat, and also the inner resistance of the battery cells creates heat as current moves through them. Dual-source heating accelerates temperature rise and guarantees thermal uniformity throughout the whole battery pack. The heating sheets aren't all identical to improve uniformity – the ones at the edges as well as the center of the packet have different resistance values to ensure equal distribution of heat despite variations in thermal dynamics. Additionally, discharging rather than charging eliminates the danger of lithium plating, a destructive and common event when batteries are charged at lower temperatures. The system can safely start heating at extremely low ambient temperatures without losing battery life or safety.

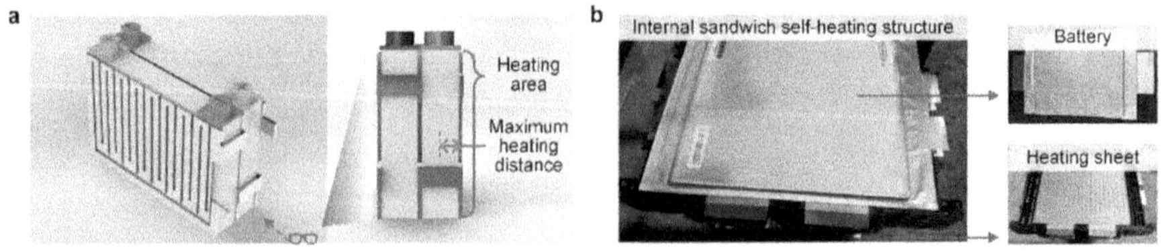

FIGURE 5.2 The schematic diagram (a) and photos (b) of the sandwich self-heating structure (Lin et al., 2023).

5.3.2 Marine Applications

In the marine sector, hydrogen fuel cells could be a renewable and green replacement for fossil fuels, with substantial benefits like zero emissions, low noise, and maintenance demands. The work of Sürer and Arat reveals that hydrogen fuel cells, particularly PEMFCs, are being increasingly implemented in marine vessels across multiple countries and projects, which include submarines and ferries in addition to tankers. A significant problem in marine applications is ensuring strong cold start ability in harsh oceanic environments. Self-heating by load cycling turns into a practical solution in this situation. At a cold start, whenever fuel cell elements are under operating temperature – potentially sub-zero – a controlled power load triggers internal resistive heating by electrochemical response. Internal self-heating gradually raises cell temperature without external heaters, stopping membrane dehydration and performance loss. The mechanism is in sync with present PEMFC cold start methods, crucial in marine applications because of limited space constraints and environmental constraints, though it is not elaborated in great technical detail in their study. Load cycling integrates self-heating to allow quicker start-up and increased operational resilience in very cold environments and contributes to cleaner maritime transportation by enhancing the practical viability of hydrogen fuel cells. The Energy Observer is the very first ship on the planet to be hydrogen-powered and hydrogen-generated. Energy Observer features a PEM water electrolyzer (PEMWE) aboard that is fueled by a solar panel and windmill to create hydrogen. The PEM drinking water electrolyzer utilizes pure seawater to create hydrogen. Several Toyota fuel cells compress and hold the hydrogen, which is utilized to drive the ship. The Energy Observer vessel system is schematically demonstrated in Figure 5.3 (Sürer & Arat, 2022).

5.3.3 Aerospace and Space Applications

Self-heating via load cycling for aerospace and space is vital for dependable start-up of hydrogen fuel cell systems at very low temperatures, where conventional heating methods are impractical or energy inefficient. Cold start ability is crucial for mission success in locations including high-altitude unmanned aerial vehicles (UAVs), satellites, orbital platforms, or space rovers, where ambient temperatures can drop considerably below freezing.

Load cycling is the intentional application of controlled power loads on the fuel cell stack to make Joule (resistive) heat internally. The technique raises the temperature of the MEA and its components from inside, stopping ice formation, improving proton conductivity, and accelerating electrochemical reactions without using bulky external heating methods. This internal heating strategy is an extremely integrated thermal management technique in space-constrained and weight-sensitive platforms like satellite power systems or space-based backup power units, lowering complexity, volume, and mass. Load cycling could be precisely controlled using onboard control electronics during power-up or standby modes in aerospace operations where autonomous system management matters most. The fuel cell system is created to be more resilient during thermal cycling resulting from orbital transitions (e.g., from shadow to sunlight) by ensuring quick response times and lessening parasitic power losses.

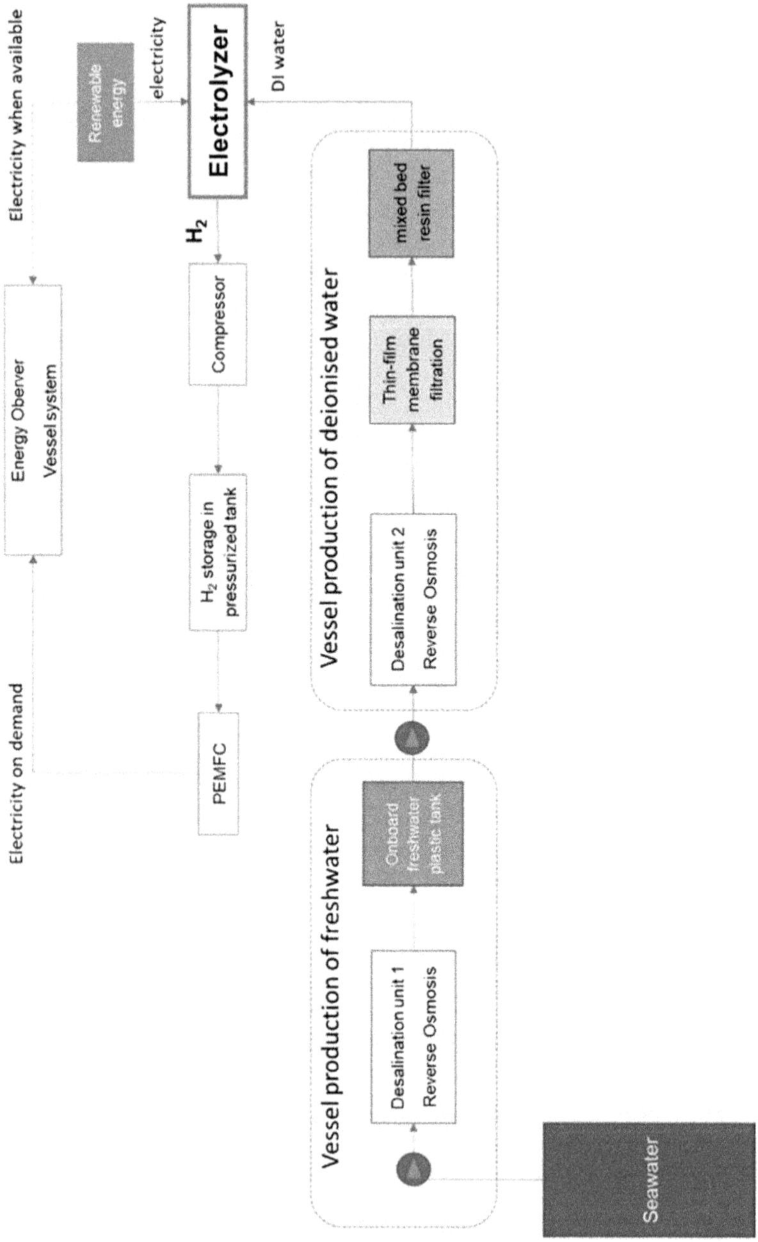

FIGURE 5.3 Schematic of energy observer vessel setup (Sürer & Arat, 2022).

Self-heating in aerospace and space platforms through load cycling enhances system reliability, extends mission times, and contributes to the general objective of deploying cold-start-capable, energy-efficient, and lightweight hydrogen fuel cell systems in hostile extraterrestrial environments.

5.3.3.1 Case Study 1: NASA – Regenerative Fuel Cell Systems for Lunar Missions

Regenerative fuel cell (RFC) systems have been thoroughly researched by NASA as a crucial source of energy production as well as storage for lunar missions, especially to power environments, landers, and research stations during the harsh lunar night, where temperatures can drop below −150°C over extended periods. PEMFCs must be started as well as operated reliably in cryogenic environments; traditional thermal management methods, including resistive or radioisotope heating, add mass and energy usage, making dependable start-up and constant operation a significant challenge. NASA examined a self-heat method employing load cycling, along with pulsed current methods, to create internal heating through the octave of the fuel cell stack. Internal heat raises the MEA temperature, stays away from ice formation, maintains proton conductivity, and also restores catalyst activity – with no external heaters. Testbed tests conducted on the ground under lunar vacuum and thermal conditions confirmed that current pulses of high frequency could effectively and efficiently increase stack temperature from sub-zero levels to operational levels. This strategy saved substantial energy as compared with traditional heaters, reduced the necessity for extra thermal components, and increased cold start reliability. Moreover, it contributed to lowering mass penalties and reducing thermal pressure on the membrane, thereby extending fuel cell lifespan. This technique was validated by NASA and can be especially relevant to upcoming Artemis missions, moon base camps, Gateway modules, and ISRU systems, where strong thermal solutions and autonomous operation are vital for cold start functionality in deep space environments.

5.3.3.2 Case Study 2: DLR (German Aerospace Center) – Horizon H2 UAV Platform

The German Aerospace Centre (DLR) has been a leader in moving forward with hydrogen-powered high-altitude UAVs, especially through its involvement in the Horizon H2 platform. The Horizon UAV is mainly powered by PEMFCs due to their high energy density and zero-emissions profile, and was created for long endurance, zero-emission flight missions like atmospheric research and surveillance, as well as communications. UAVs flying over 10,000 meters encounter serious operational difficulties because of the intense thermal environment at cruising altitude, where the ambient air temperature can drop below –40°C, threatening the activation as well as longevity of fuel cells. The Horizon H_2 platform adopted a self-heating method that involved carefully modulated electric loads applied to the fuel cell stack during the first start-up to overcome the extreme cold start conditions. This technique induced Joule (ohmic) heating inside the MEA, slowly increasing the internal temperature of the stack from sub-zero conditions to its operating range. This internal load-induced heating technique permitted autonomous cold starts even under freezing and low-pressure temperatures, unlike traditional external heating systems, which are generally heavy, complex, and power-intensive. The procedure decreased thermal gradients in the cell, decreasing membrane stress and the possibility of physical failure because of ice. This load cycling strategy employed by DLR in a light UAV platform was notable because it removed the need for extra thermal controller subsystems that are not doable in mass- and volume-constrained aerospace designs. The Horizon H_2 UAV would be able to handle high-altitude missions with enhanced energy efficiency, quicker start-up times, and enhanced fuel cell resilience, leading to these changes. This case shows precisely how self-heating via load cycle is a practical thermal management technique, along with a crucial enabler, for high-reliability, cold-weather hydrogen fuel cell operation in challenging aerospace applications.

5.3.3.3 Case Study 3: ESA (European Space Agency) – Fuel Cells for Satellite Backup Power

PEMFCs are examined by the European Space Agency (ESA) as a possible auxiliary or backup power supply for spacecraft and orbital platforms, particularly during long-duration missions where energy control as well as thermal redundancy are crucial. Satellites in low Earth orbit (LEO) or geostationary orbits frequently go through eclipse phases – times when the spacecraft passes through the Earth's shadow and also loses immediate access to solar energy. The internal temperature of the spacecraft could reduce considerably during these shadowed intervals, often to cryogenic or sub-zero conditions, placing great stress on onboard energy systems, particularly fuel cells, which depend on hydrated membranes and active catalysis.

ESA has examined the utilization of self-heating through electrical load cycling to pre-warm the MEA of POMFCs during cool phases to deal with this particular issue. The method involves emitting short, precisely controlled current pulses throughout the fuel cell stack to cause Joule heating, which gradually raises the heat inside the active parts without the need for costly, bulky external heaters. This approach is especially advantageous in space applications, in which mass, volume, and power budgets are severely constrictive and where autonomous operation is important due to limited intervention capabilities. ESA conducted specialized studies and system simulations, which demonstrated that the internal heating technique considerably improves cold start reliability when changing out of the eclipse (cold) to the sunlit (power demand) stage. The strategy of pre-heating the MEA just before energy generation is needed to stop ice formation, membrane dehydrations, and sluggish protonic conduction, all of which are typical cold start issues in space fuel cells. Moreover, the approach reduces thermal shock and structural pressure on the membrane and catalyst layers, thereby enhancing stack longevity and operational stability across multiple orbital cycles.

5.4 DESIGN CONSIDERATIONS FOR LOAD CYCLING SYSTEMS

Load cycling is the area of designing that needs the careful consideration of the heat generation, the systems integration, and energy efficiency of designing self-heating solutions to hydrogen fuel cell chilly starts. Cyclic loading and operating the fuel cell at higher current densities may soon rectify the internal temperatures, permit faster cold starts using the very expansion of heat; however, that ought to be combined to avoid heavy energy draw and potential wear. Further-developed approaches mix self-training with external approaches, such as liquid or perhaps coolant pre-heating, phase change materials, or perhaps thermochemical metal hydride pre-heaters, to accomplish improved efficiency and greater speed, with lower external energy consumption and ice production hazards. Dynamic control algorithms that alter the airflow and load on the fly may also provide improved reliability of cold start and temperature uniformity around −40°C. The most recent has involved the development of hydrogen storage systems that emit heat during the absorption of hydrogen, reducing cooldown times and energy requirements by drastic margins. The AHP technology offers rapid and water-free ohmic heating that is accompanied by minimal wreckage, suggesting that it can substitute the already effective and potent cold start applications that are used in extreme cold temperatures (Zhu et al., 2024). To ensure safe, reliable, and improved application of the method, some key design issues ought to be addressed across the electrochemical, electrical, thermal, and system integration fronts.

5.4.1 Load Cycling Strategy

Load cycling strategy is an important designed aspect in cold start systems with self-heating hydrogen fuel cell, as this allows the internal Joule heating length of the MEA to be warmed up without any external thermal gadgets. The method would be to inject and withdraw well-controlled large current pulses into the fuel cell stack and relies a lot on optimization of pulse frequency, amplitude, and duty cycle. Their rate of heat production is influenced directly by pulse frequency, amplitude of these pulses, with higher exciting pulses creating more heat but potentially leading to overheating in local areas or dry-out of the membrane, whereas the lesser intense pulses may not create enough heat to increase the temperature of the stack rapidly enough to counteract freezing. Equally, the pulse frequency should be rather balanced, being too fast cycling can cause thermal stress and inefficiency, and slower cycling can cause a long start-up time. Heating is continuous to the degree that the ratio of the time the heating is effective to the idle time in each heating cycle, known as the duty cycle. Low duty cycles have the advantage of increasing temperatures fast, but a disadvantage of poor heat distribution and possible material fatigue, whereas higher duty cycles are faster in heating and uniform in their thermal gradient, but will cost you in the start-up time. In order to control this complexity, a knowledgeable control algorithm is therefore vital; it has to observe the actual temperature data given by temperature sensors, current levels, and voltage combinations and modify the pulse profile at that point in time to avoid hot spots, guarantee temperature uniformity, and operational safety. Such algorithms may include stack thermal models for internal temperature gradient forecasting and pulse pattern setting in anticipation. These control strategies may be combined with the general power management system in more advanced fuel cell systems – as in vehicles, UAVs, or satellites – and ensure the smooth adjustment of heating needs to prevailing energy sources. Finally, a well-designed load cycling strategy enables the fuel cell to start reliably in sub-zero temperatures with minimum power consumption, while preserving system integrity and extending system life.

5.4.2 Membrane Electrode Assembly (MEA) Durability

A very important consideration for the self-healing hydrogen fuel cell cold start system design is the reliability and durability of MEA under repeated freeze–thaw cycles. The MEA is stressed thermally and mechanically during cold starts, and freeze–thaw tolerance is an important parameter. Below 0°C, residual water in the membrane or GDL may freeze and expand in volume, pressing down on the polymer structure. When this expansion–contraction cycle is not properly managed, the membrane may suffer mechanical cracking, delamination of the catalyst layer, or pinhole formation. To counter this, MEAs must be constructed of sufficiently flexible and mechanically resilient materials – reinforced composite membranes or specially treated ionomers – to withstand the stresses of ice formation and thawing without permanent damage. Important too is hydration management, since proton conductivity in the membrane depends on water. A dry membrane cannot conduct protons effectively, causing voltage losses or start-up failure, and excess water can freeze, causing structural damage. This means that optimal hydration during a cold start is very important. This involves controlling humidity, hydrogen flow rates, and thermal gradients, and using membranes that do not lose water excessively or freeze under low temperatures. Catalyst stability under load cycling conditions is also important. Repeated pulses during self-heating may induce local fluctuations in potential that may cause platinum (Pt) to dissolve, migrate, and agglomerate in high-surface area catalyst layers. This reduces electrochemically active surface area and may also result in platinum band formation in the membrane, causing membrane thinning and electrical shorting. To prevent such degradation, robust catalyst formulations are required, for example, Pt-alloy catalysts or nanoparticles based on corrosion-resistant carbon or conductor metal oxides. In addition, stabilizing agents or an optimized ionomer–catalyst interface can reduce mechanical and electrochemical stress during freeze and load cycling reactions.

5.4.3 Thermal Management

For a hydrogen fuel cell cold start self-heating strategy, a well-designed thermal management system is required to meet the localized heating distribution, thermal insulation, and temperature-monitoring challenges. If the heat is generated internally via load cycling – by applying electrical current to the fuel cell stack – then the heating effect is not uniform, at least not in long stacks or stacks with uneven gas distributions. In such cases, the inlet sections where hydrogen and air enter the cells first may heat up faster because of higher reactant concentrations/lower thermal inertia, whereas the outlet sections are colder. Such thermal gradients across the MEA delay full-stack activation and introduce mechanical stress that damages the membrane or catalyst layers over a long time. To counter this, the stack architecture should be optimized for even thermal distribution. It could be as simple as segmented flow fields, uniform reactant delivery, or even thermal spreaders with highly conductive bipolar plate materials to disperse heat more evenly over the active area. Equally important is the thermal insulation around the stack enclosure. Most heat generated during load cycling is lost to the environment without proper insulation, especially at sub-zero ambient temperatures. This lowers the temperature rise rate and increases the energy needed to reach operational thresholds. Insulating materials like aerogels, vacuum panels, or multi-layer thermal foils should be light in weight but efficient in reducing conductive, convective, and radiative heat losses. Also, the insulation should be compact and tough under vibration and thermal cycling for mobile or aerospace applications with limited space and weight. For safe and precise heating control, the system must have accurate and strategically placed temperature sensors. These sensors give thermal feedback from the inlet/outlet, center, and bipolar plates to monitor heat propagation during self-heating. This information is important to the adaptive control algorithms that regulate current pulses to avoid thermal runaway, uneven heating, or overcompensation, leading to membrane drying or cracking. Furthermore, these sensors are very useful for early fault detection, for example, safety shutoffs or switching to backup systems when temperature profiles are detected abnormally. Together, they ensure that self-heating via load cycling is not only effective in raising the stack temperature but also thermally balanced, energy-efficient, and structurally safe for repeated cold start operations in hydrogen fuel cell systems.

5.4.4 Electrical and Power Electronics Design

A critical subsystem for self-heating via load cycling hydrogen fuel cell cold start system is the electrical subsystem design, including the current control interface, power supply compatibility, and voltage regulation mechanisms. The key to the load cycling approach is to apply short-duration high-frequency pulses to the fuel cell stack to cause internal Joule heating. This needs a current control interface that can dynamically modulate pulse width, frequency, and amplitude online. Such an interface should provide accurate and responsive current control but also provide electrical isolation, low latency, and high efficiency to avoid energy loss or electromagnetic interference. Advanced designs may use pulse width modulation (PWM), digital signal processors/embedded microcontrollers programmable to vary current with respect to real-time temperature feedback/stack operating conditions. Also, the interface must be robust enough for rapid switching and thermal cycling without compromising performance or introducing instability in the electrical system. Also important is power supply compatibility. Energy required for load cycling during cold start comes from either onboard batteries or APUs. Hence, the load cycling system should be synchronized with the available power source in voltage range, current capacity, and timing. In automotive applications, this may mean drawing energy from the 12V or 48V battery system, while power management in microgravity and under tight energy budgets is required for aerospace or satellite platforms. Power supply should deliver consistent current pulses without triggering overcurrent protection or voltage drops elsewhere in the system. Voltage stability during pulse application is also an important issue for fuel cell components protection and system integrity. Speedy and repeated current loading can cause voltage oscillations that are unregulated and may cause membrane dehydration or cathode flooding. Dehydration leads to proton conductivity loss and membrane cracking, and excess water at the anode may prevent oxygen transport and reduce reaction yield. To limit these risks, the system must contain real-time voltage monitoring and regulation circuits to stabilize the applied potential across the stack during each load pulse. This may include DC-DC converters, active feedback loops, and failsafe mechanisms to shut down/adjust the load upon unstable voltage behavior. Together, these electrical control elements assure that load cycling provides safe and efficient self-heating to enable fast cold start of hydrogen fuel cells without jeopardizing component health or operation continuity.

5.4.5 Hydrogen and Water Management

Good hydrogen and water management is a crucial design consideration when self-heating through load cycling is implemented in hydrogen fuel cell cold start solutions. During cold start, the internal temperature of the fuel cell stack should be increased without producing additional water that freezes and causes severe mechanical damage or blocks reactant flow paths. To deal with this, a no-production water strategy is often used. In this approach, electrical load pulses are applied in a nonreactive or open circuit condition, where current is drawn to produce internal heat by ohmic resistance, but the electrochemical reaction of hydrogen with oxygen is suppressed. This prevents product water from forming at the cathode, which otherwise would freeze in sub-zero conditions and damage the MEAs, GDLs, or flow channels. Besides, uniform heating across the stack requires precise hydrogen distribution management. Uneven hydrogen flow in load cycling can cause localized gas starvation/flooding. Any portion of the stack that receives more hydrogen than others may heat up faster, causing thermal gradients that may stress the membrane and catalyst layers. In turn, insufficient hydrogen in certain regions causes increased resistance, uneven current distribution, and poor heat generation. To avoid such imbalances, the hydrogen delivery system must be calibrated often with flow field designs, backpressure regulators, and humidification control to ensure even distribution of hydrogen over the stack of cells. This is particularly important during start-up when low temperature increases gas viscosity and decreases diffusion rates, causing flow control to be challenging. Besides, residual water from previous shutdown cycles may be trapped in the MEA or flow fields. Clear or redirect this water using passive water removal techniques, stack orientation, or integrated drainage features before load cycling starts. Advanced applications – for example, automotive or aerospace – use closed-loop control systems that actively monitor hydrogen flow, temperature profiles, and water content for optimal self-heating without stack integrity degradation.

5.4.6 Stack and System Configuration

An intelligent and efficient stack architecture can be an important contribution to optimizing self-heating via load cycling for hydrogen fuel cell cold start solutions using segmented stack heating and internal bypass paths. In traditional full-stack heating approaches, energy is injected uniformly across all cells, irrespective of whether every section requires the same thermal input. Yet, it can result in excess energy consumption, especially for long or high-power stacks in automotive or stationary systems. Segmented stack heat provides a better energy efficiency solution by selectively heating critical regions like the inlet cells, which are usually the coldest during start-up and most susceptible to freezing due to initial exposure to cold hydrogen and air. Many of these regions experience ice formation and water condensation, resulting in local cold spots and increased resistance. When applying load pulses in these zones – via electrical segmentation or targeted power distribution – the system can recover thermally when needed most and save start-up time and energy consumption while preventing localized freeze damage of the MEA.

Besides, stack design, including internal gas bypass paths, provides better thermal uniformity during load cycling. These bypass channels are incorporated in the stack flow field/manifold system to redirect the flow of reactant gases, such as hydrogen and air, to distribute heat from pulsed current application to other regions. When applied current produces ohmic heat in some sections, warmed gases are circulated – passively by convection or actively by controlled flow – into cooler regions of the stack. This allows temperature equalization and reduces thermal stress due to steep gradients across adjacent cells. Some advanced designs use dual flow field geometries or looped configurations to encourage bidirectional gas movement, thereby enhancing heat redistribution. In addition to their function as safety or purge channels, bypass paths may let trapped condensate or excess water vapor escape before it freezes. Segmented heating and bypass architectures together provide a strong synergy effect – faster, safer, and more energy-efficient cold starts when localized heating as well as system-wide thermal management are addressed. All these design innovations are very useful in applications with tight energy budgets or extreme environmental conditions like aerospace platforms, remote backup systems, or next-generation electric vehicles.

5.4.7 Material Compatibility

For cold start operation of hydrogen fuel cells with self-heating via load cycling, the mechanical and material integrity of the bipolar plates, GDLs, and compression hardware is of paramount importance. Cold starts involve repeated thermal cycling from sub-zero temperatures to operational levels and back – this places great stress on internal components. These bipolar plates must conduct electricity, disperse gases, and conduct heat and water without warping, corroding, or losing conductivity. Preferred materials are carbon-based plates having good corrosion resistance and thermal stability, and coated stainless steel plates having mechanical robustness and good conductivity of current. These plates are usually coated with gold, chromium oxide, or carbon composites, resisting degradation in a humid acidic fuel cell environment and reducing the interfacial contact resistivity to maintain stable current flow in dynamic thermal cycles. Also, the GDLs between the catalyst layer and the bipolar plate must not lose their porosity, structural integrity, and hydrophobic properties with increasing temperature. Carbon paper or carbon cloth, PTFE-treated, are typical materials; They must, however, tolerate expansion and contraction stresses without cracking or becoming brittle, because any structural failure would block gas flow, block water removal, and create hot spots that increase membrane damage risk during cold starts. Also important is the design of the stack's compression hardware, which must exert a uniform and precisely calibrated clamping force across all cells to accommodate thermal expansion and contraction. Uneven pressure distribution may cause mechanical stress, seal leakage, or delamination of the MEA. To counter this, spring-loaded bolts, Belleville washers, or torque-controlled fasteners maintain constant compression during thermal cycling in advanced clamping systems. End plates and frame materials also must be thermally stable and expansion compatible with the remainder of the stack. Together, they ensure that the stack is mechanically sound, thermally resilient, and electrochemically efficient during and after cold start cycles, enabling tough performance in demanding automotive, aerospace, and portable fuel cell applications.

5.4.8 Energy Efficiency and DOE Targets

Designing self-heating strategies by load cycling for hydrogen fuel cell cold start solutions is an important step in considering start-up energy budget and start-up time requirements as per industry standards or application requirements. In cold climates, system responsiveness, performance degradation, and user experience are critical goals – especially in automotive fuel cells, where driver expectations require near-instant power availability. Regulators and technology performance benchmarks like those defined by the US Department of Energy (DOE) require a PEM fuel cell system to start from 0°C ambient temperature with less than 5 MJ total Energy. That total includes all energy used by auxiliary systems; therefore, the load cycling-based heating design must be very efficient – producing enough heat to bring the MEA temperature up to the operational range while drawing minimum electrical energy from the vehicle battery or power management unit. This requires engineers to optimize the current pulse profile, duty cycle, thermal insulation, and power electronics of the self-heating process. Meanwhile, start-up time targets must be met – usually 30–120 s for passenger vehicles. This time span is balanced between user convenience and system safety, so the stack can reach the required hydration and thermal stability for the current generation without ice formation or mechanical damage. In aerospace and stationary applications where rapid response is less critical, the energy budget may be squeezed even more by onboard power constraints or long-duration energy planning. Therefore, rapid heating, energy efficiency, and component longevity must be optimally balanced. Integrating real-time temperature feedback, intelligent current modulation, and thermal insulation may enhance start-up kinetics and minimize energy waste. By considering these targets in design, self-heating load cycling systems may meet or exceed application performance requirements, allowing fuel cell technologies to operate reliably in a wide range of environments and use cases.

5.4.9 Safety and Control Logic

Robust safety and control mechanisms are required when self-heating via a load cycling system for hydrogen fuel cells cold starts, in particular, overcurrent protection, start-up failure detection, and system integration with higher-level energy management are to be considered. Overcurrent protection prevents thermal overloading/ electrical faults from occurring by preventing excessive current application to the stack in pulsed heating. Without this safeguard, localized overheating will damage the MEA, degrade catalyst layers, or even produce electrical shorts under fluctuating cold start conditions. More reliable start-up failure detection logic must be embedded in the control system, which will continuously monitor stack temperature trends and pulse efficiency. If the system senses inadequate temperature rise within a predefined window – for example, ice blockage, membrane dehydration, or component failure – then it should be able to abort the self-heating sequence/switch to a backup heating method to avoid stack damage and keep operation. Finally, the effective self-heating system should be fully integrated with a battery management system and an energy management system (EMS) for coordinated control of power flow/temperature monitoring/safety interlocks. Cette integration enables real-time communication between subsystems and adaptive load cycling based on available battery capacity, external ambient conditions, and system thermal load while respecting operational budget. Together, these features form a smart, fail-safe, and application-aware heating strategy required for cold start reliability in automotive, aerospace, and backup hydrogen fuel cell systems.

5.5 MONITORING AND CONTROL IN SELF-HEATING SYSTEMS

Good monitoring and control are essential for self-heating via load cycling to achieve a cold start in hydrogen fuel cells under conditions where fuel cell performance is severely impaired by sub-zero temperatures. This method works on precise pulsed current application to produce internal heat by Joule heating and requires a very precise feedback control to ensure the safe and effective heating process. The core of the monitoring system consists of temperature sensors – usually thermocouples/RTDs – placed along the stack in inlet/outlet zones, mid-stack locations, and bipolar plate interface locations. These sensors provide real-time thermal data to the control algorithm for dynamic adjustment of current pulse parameters like amplitude, frequency, and duty cycle. This feedback assures uniform heating over the entire MEA and avoids thermal gradients, which may lead to membrane stress or denaturation. This is usually controlled by a dedicated microcontroller or embedded processor that runs a smart thermal management algorithm. Using predictive modeling, real-time feedback, and historical data to select the most efficient heating strategy. It may, for example, increase pulse frequency at initial start-up when the stack is cold and taper pulses as temperature approaches operational thresholds. In advanced systems, thermal behavior may be predicted using model predictive control (MPC) or fuzzy logic (Wu et al., 2020).

The control system must also monitor voltage and current behavior during load cycling, besides temperature regulation. Changes in stack voltage could reflect membrane dehydration, local flooding, or electrical faults. When anomalies are detected, the system can issue corrective commands like pulse intensity reduction or cooldown start, and log diagnostic data for further evaluation. When start-up failures occur, for example, insufficient temperature rise or inconsistent voltage response, the system can initiate failsafe protocols such as switching to backup heating methods or aborting start-up completely to protect stack components. Integration with the vehicle Battery Management System/Energy Management System/EMS is also necessary. This allows coordinated energy allocation during initial power-up if battery demand is high. Communication to BMS/EMS ensures that load cycling does not affect overall system energy balance or interfere with other start-up processes (Xu et al., 2022). New strategies to monitor and control self-heating during cold starts are suggested, including load cycling. The AHP method, with no water production, cycles hydrogen flow to produce ohmic heating in just 30 s, and shows little degradation over thousands of cycles, thus enabling fast and easy self-heating of a PEMFC starting from −30°C in just 30 s. Combining load cycling and optimized power control sequences improves heating efficiency and start-up speed. Some other innovative solutions are coupling PEMFCs with metal hydride storage systems, which heat the stack by exothermic hydrogen absorption, reducing cold start time and energy consumption. Combining internal self-heating and external pre-heating, like liquid heating or catalytic heaters, also improves cold start performance and energy efficiency. Simulation studies show that hydrogen flow arrangements and starting strategies like lower start-up voltages/counter-flow configurations can reduce ice formation and speed warm-up (Xu et al., 2023).

5.6 ENERGY EFFICIENCY AND HEAT MANAGEMENT

Optimal performance and long life of hydrogen fuel cells are dependent on energy efficiency and heat management during cold start conditions. Load cycling – modulating the electrical load to produce self-heating may quickly raise cell temperatures, shorten cold start times, and reduce performance losses due to low temperatures. Advanced thermal management strategies such as phase change materials, microchannel heat exchangers, and liquid cooling loops are investigated to regulate temperature fluctuations and heat dissipation in PEM and solid oxide fuel cells for better performance and system life. Integrating predictive thermal modeling with AI-driven control algorithms optimizes the real-time heat distribution to achieve adaptive cooling and load sharing. Heat recovery is another important strategy: Up to 60% of the hydrogen energy content is lost as heat that can be recovered for combined heating, cooling, and power applications or used to heat reactants/improve cold start performances. Energy management strategies focusing on heat utilization – peak clipping/load reduction – could ensure almost all generated heat is used, thereby reducing waste and increasing system efficiency – even in off-grid/variable renewable scenarios. Flexible load regulation/demand management may also reduce hydrogen consumption/thermal storage needs, thereby improving exergy efficiency/emissions. An advanced system can apply an adaptive control algorithm for heating modulation based on continuous feedback from temperature, current, and voltage sensors. Those systems can reduce load intensity when the stack reaches its working temperature to prevent heat buildup and improve long-term durability. Also, energy-efficient self-heating allows a short start-up time of 30–120 s without compromising MEA integrity, catalyst degradation, and fuel cell component life extension (Zhao et al., 2024).

5.7 INNOVATIONS IN LOAD CYCLING AND SELF-HEATING

Ice accumulation and mechanical stress are the major challenges for cold start in polymer electrolyte membrane fuel cells – PEMFCs – at sub-zero temperatures. Innovative load cycling, self-heating strategies for fast and controlled internal heat generation are emerging solutions to this challenge. This approach makes use of the Joule heating effect in transient high-current operation to raise the stack's internal temperature above the freezing point of water to accelerate ice melting and facilitate start-up. Recent developments involve adaptive load cycling techniques in which controlled current pulses are applied in short intervals to produce localized heating without membrane dehydration or catalyst degradation risk. They are now dynamically modulated in shape, frequency, and amplitude with real-time sensor feedback for energy efficiency and thermal uniformity in the cell. Such techniques allowed localized thaw without external heaters and thus reduced parasitic power consumption and improved system compactness. Thermally coupled systems consisting of metal hydride tanks and PEMFC use the exothermic heat of hydrogen absorption to pre-heat the stack, reducing cold start time by almost 60% and energy consumption by over 60% compared with conventional methods. Another innovation is the combination of thermal sensors with intelligent control algorithms based on MPC or AI decision systems (Melnik et al., 2024). Such systems continuously monitor critical parameters such as internal resistance, temperature gradients, and voltage response to fine-tune the cycling protocol. In addition, novel materials with improved thermal conductivity and tailored ice-cracking thresholds are employed in the catalyst layer and MEA to improve the durability during repeated thermal cycles. Newer designs also consider hybrid heating mechanisms combining load cycling with electrical or catalytic pre-heating to produce a robust and scalable cold start under extremely low temperatures (below –30°C). Hybrid approaches combining internal self-heating, externally assisted pre-heating, and phase change materials further optimize energy use and speed, liquid heating plus self-starting, reducing start time and electric heating requirements.

5.8 CURRENT STATE OF IMPLEMENTATION AND PROGRESS IN SELF-HEATING THROUGH LOAD CYCLING

Implementation of self-heating by load cycling in hydrogen fuel cells cold start strategies has gone from conceptual demonstrations to practiced, system-integrated approaches presently. Newer fuel cell systems incorporate load cycling protocols to provide sufficient internal heat via controlled current pulses for start-up at –20°C ambient conditions. Some automotive OEMs and research institutions have validated this approach on single-cell and stack levels and found it to be effective in thawing frozen membranes without excessive parasitic energy loss. Advancement of control strategies such as thermal sensors, AI-based feedback loops, and predictive algorithms allowed precise regulation of pulse magnitude and duration to avoid localized overheating/dry-out damage. Commercial prototypes now incorporate real-time energy management units synchronizing load cycling with other auxiliary subsystems as air supply and humidification. Besides, integration with hybrid heating methods, including a combination of load cycling with low-power electrical heaters or exothermic reactions, is reported to increase the reliability in colder climates (below –30°C). Further developments of robust MEAs for better ice-tolerance and thermal cycling resistance are in line with this progress. Research is ongoing to improve uniform heat distribution, start-up times below 30 s, and reduce voltage losses during cycling.

Cold-weather fleets and portable power units are being tested in field trials against traditional cold start methods. The tech is getting into commercial vehicle applications such as buses, delivery vans, and off-grid systems. Key patents such as US 8,715,873 B2 describe controlled load cycling to induce internal heating during sub-zero start-ups. That patent is active through 2030 and is one of the earliest demonstrations of pulse-based cold start strategies. In recent patents like US 12,087,980 B2 and US 11,349,138 B2, energy flow regulation during start-up using thermal feedback is proposed in smart control systems for system reliability improvement. In US 10,249,890 B2, an approach to redirect early power output to warm auxiliary components instead of drive systems is presented with practical application readiness. Some emerging applications – WO 2020/173166 A1 and US 2025/0023074 A1 – demonstrate a shift to hybrid control approaches – combining electrical, resistive, and catalytic heating methods – although these are not yet examined.

5.9 CONCLUSIONS

Self-heating by load cycling is a promising approach to cold start problems of hydrogen fuel cells, especially in sub-zero conditions. Utilizing internal heat generation mechanisms like Joule heating, electrochemical activation losses, double-layer charging effects, hysteresis, and microstructural friction enables fast and energy-saving start-up without external heating systems. Integration of advanced materials, intelligent control algorithms, and hybrid thermal management techniques allowed cold start time reductions of up to 59% and energy savings of over 60%. This chapter describes how load cycling exploits several internal pathways to raise the temperature uniformly to avoid ice formation and maintain proton conductivity. Besides, the new membrane, catalyst layer, and GDL design have given the material good performance under cyclic thermal and mechanical stress. Practical examples in automotive, aerospace, marine, and backup power systems show that this technology is adapted to operational environments. Parallel sophisticated monitoring systems with adaptive control logic optimize pulse profile and thermal uniformity to prevent degradation of components. Adding predictive modeling, smart materials, and hybrid heating systems will further refine self-heating strategies. Self-heating via load cycling is a key enabler of fuel cell operation as the hydrogen economy develops, especially in cold climates. It addresses the interplay of thermal management, materials engineering, and system-level integration for a new generation of high-performance, durable, and cold startable hydrogen fuel cell systems.

REFERENCES

Jung, H., Noh, J., & Um, S. (2011). Experimental study of electrical switching characteristics of vanadium oxide thin films on bipolar plates for improving thaw-at-start. In *International Conference on Fuel Cell Science, Engineering and Technology* (Vol. 54693, pp. 293–299).

Lin, C., Kong, W., Tian, Y., Mao, Y., Zhou, E., Shao, Q., Wu, N., Liu, J., Yu, X., & Huang, S. (2023). Sandwich self-heating structure-based lithium-ion battery system and its application in the fuel cell bus for Beijing Winter Olympic Games. *Energy Conversion and Management*, 284(April), 116977. https://doi.org/10.1016/j.enconman.2023.116977

Liu, H., Si, D., Ding, H., Wang, S., Zhang, J., & Liu, Y. (2021). Cold start capability and durability of electrospun catalyst layer for proton exchange membrane fuel cell. *International Journal of Hydrogen Energy*, 46(19), 11140–11149. https://doi.org/10.1016/j.ijhydene.2020.06.032

Melnik, D., Bürger, I., Mitzel, J., Käß, J., Sarkezi-Selsky, P., Jahnke, T., & Knöri, T. (2024). Energy efficient cold start of a polymer electrolyte membrane fuel cell coupled to a thermochemical metal hydride preheater. *Applied Energy*, 359, 122585.

Meng, K., Chen, B., Zhou, H., Shen, J., Shen, Z., & Tu, Z. (2022). Investigation on degradation mechanism of hydrogen–oxygen proton exchange membrane fuel cell under current cyclic loading. *Energy*, *242*, 123045. https://doi.org/10.1016/j.energy.2021.123045

Sürer, M. G., & Arat, H. T. (2022). Advancements and current technologies on hydrogen fuel cell applications for marine vehicles. *International Journal of Hydrogen Energy*, 47(45), 19865–19875. https://doi.org/10.1016/j.ijhydene.2021.12.251

Wang, G., Utaka, Y., & Wang, S. (2021). Improvement of cold start performance of polymer electrolyte fuel cell using microporous and gas diffusion layers with patterned wettability. *International Journal of Energy Research*, 45(9), 13760–13770. https://doi.org/10.1002/er.6704

Wen, J., Si, D., Wang, S., Ding, H., Li, C., Ono, K., Chen, Z., & Zhang, J. (2019). Alternate hydrogen pump method enables start-up from −30°C for graphite-bipolar-plate proton exchange membrane fuel cells. *Journal of the Electrochemical Society*, 166(14), F1112–F1116. https://doi.org/10.1149/2.0091914jes

Wu, K., Xie, X., Wang, B., Yang, Z., Du, Q., Xuan, J., Zu, B., Liu, Z., & Jiao, K. (2020). Two-dimensional simulation of cold start processes for proton exchange membrane fuel cell with different hydrogen flow arrangements. *International Journal of Hydrogen Energy*, 45(35), 17795–17812. https://doi.org/10.1016/j.ijhydene.2020.04.187

Xu, H., Wang, S., Si, D., & Zhang, J. (2022). Cold start of PEMFC using alternating hydrogen pump: Part I. Mechanistic modeling. *Journal of the Electrochemical Society*, 169(5), 54527. https://doi.org/10.1149/1945-7111/ac6f86

Xu, H., Wang, S., & Zhang, J. (2023). Cold start of a PEMFC using alternating hydrogen pump: Part II. Optimization and comparison with rapid self-heating. *Journal of the Electrochemical Society*, 170(2), 24511. https://doi.org/10.1149/1945-7111/acbb33

Zhao, J., Tu, Z., & Chan, S. H. (2024). Energy management strategy for accurate thermal scheduling of the combined hydrogen, heating, and power system based on PV power supply. *International Journal of Hydrogen Energy*, 78(January), 721–730. https://doi.org/10.1016/j.ijhydene.2024.06.341

Zhu, W., Wang, Y., Gao, S., He, F., Li, C., & Wang, Y. (2024). Study on cold start of fuel cell stack combined with liquid heating and fuel cell self-starting. *Case Studies in Thermal Engineering*, 60(July), 104814. https://doi.org/10.1016/j.csite.2024.104814

6 Use of Hydrogen Burners

6.1 INTRODUCTION

Hydrogen burners are a kind of special developed technology that uses hydrogen as a major combustion fuel. As the world heads toward going decarbonized and implementing sustainability in industries, these burners are an important technology that can replace conventional fossil-based burners in high-temperature applications (metalworking, glass, ceramics, and power generation), among others. The main interest of the use of hydrogen is that it has clean combustion properties; unlike hydrocarbons, burning hydrogen in combustion with oxygen results only in the emission of steam, with the absence of emitting carbon dioxide (CO_2), which is one of the main factors of climate change reduction and, consequently, a significant contribution to its mitigation. However, the behavior of hydrogen under combustion is radically different from that of conventional hydrocarbon fuels, with different engineering issues. Hydrogen has a very high flame speed, low ignition energy and high diffusivity properties, which although may provide rapid ignition, maximum combustion efficiency are also likely to cause flashback (the flame propagating upstream into the burner) blow-off (extinction of the flame by excessively high flow velocities) and thermal generation of nitrogen oxides (NO_x) (formation at high flame temperatures). Such dangers demand the creation of special burner surfaces and excellent security methods. To overcome such complexities, advanced designs of burners have been introduced, such as Micromix, partially premixed, and low-NO_x designs. Micromix burners disintegrate the fuel and oxidizer into many small jets, and the fuel–oxidizer mixture is then combusted on a microscopic scale by a factor of many tens in terms of stability and local peak temperature, thus producing very low emissions of NO_x. Partially premixed systems provide an intermediate position between the premixed and diffusion flames that provides the maximum mixing of the fuel and air mixture and minimizes the danger of flashback.

Numerical modeling as well as experimental studies have been of great significance in the invention of a hydrogen burner. With the help of Computational Fluid Dynamics (CFD) models, it is possible to model the behavior of flames, heat transfer, and pollutant formation at different operating conditions, which can be used to gain insights into flame optimization of burners. This research has verified the reality that combustion efficiency can be augmented using fuels enriched with hydrogen so that flammability limits can be exponentially widened, and so that carbon dioxide emissions can be lowered on account of co-firing. Nevertheless, they also show how accurate mixing ratios, temperature fields, and burner geometries are needed to deal with NO_x emissions and as safe operation. The new hydrogen burners are being prepared to be more fuel flexible. These burners can either run in pure hydrogen,

DOI: 10.1201/9781003657866-6

natural gas-hydrogen blend, or two-fuel mode; thus, they can be used both to upgrade existing infrastructure and to realize future-proof energy systems. This versatility allows a transitioning plan at a slow pace in some areas where switching entirely to hydrogen is still not possible because of logistical shortages or financial inabilities. Although these have been technological innovations, there are other major challenges to the wide application of the hydrogen burners. These are the expense and ramp-up ability of generating hydrogen, especially green hydrogen through electrolysis; the danger of the flammable nature of hydrogen and how it is stored at very high pressures; and the insufficient infrastructure to distribute and deliver hydrogen. In addition, the shift toward hydrogen combustion can necessitate capital-intensive modification of the current industrial systems and adherence to new regulatory standards.

6.2 WORKING PRINCIPLE OF HYDROGEN BURNERS

Hydrogen-burning devices combine hydrogen gas with air to form a high-temperature flame – the challenge is controlling combustion for safety, efficiency, and low emissions. As hydrogen is very diffusive and flames are very fast, special burner designs such as Micromix and swirl burners are employed to achieve rapid and uniform mixing, prevent flashback (return of flame into the burner), and reduce the emission of NO_x. The injection of hydrogen into air in Micromix burners produces many small diffusion flames that burn safely and without flashback, and short residence times at high temperatures keep NO_x emissions low (Ayed et al., 2017). The reaction mechanism for hydrogen burners is.

$$2H_2 + O_2 \rightarrow 2H_2O + \text{Heat}$$

This reaction releases a great deal of energy as heat. Swirl burners swirl to mix and stabilize the flame to reduce pollutants and improve combustion stability. The premixing of hydrogen with air prior to combustion is an important step toward stable operation with low NO_x output, as the uniform mixture prevents hot spots with high emission potential. Advanced designs also employ recirculation zones and porous media for flame stability/heat transfer and additive manufacturing for rapid prototyping/optimization of burner components. Hydrogen combustion gives off no carbon dioxide and is therefore a clean energy option; however, the design should consider the higher NO_x formation and flashback risk of hydrocarbon burners compared to conventional hydrocarbon burners.

6.2.1 Working Principle

6.2.1.1 Step 1: Fuel and Air Supply

The fuel and oxidizer supply systems in hydrogen combustion systems are important for complete, efficient, and safe energy release. Hydrogen gas (H_2) is fed to the burner from high-pressure storage cylinders, gas pipelines from on-site reformers (e.g., steam methane reformers or water electrolysis units), and compressed hydrogen tanks. Under the extremely low molecular weights (2.016 g/mol) and high diffusion coefficients (~0.61 cm^2/s in air at STP), its flow behavior is unique compared with conventional hydrocarbon fuels. As a result, fuel is controlled by pressure regulators, flow control valves, and mass flow controllers (MFCs) to provide laminar, non-turbulent flow and prevent pulsation or pressure spikes.

The oxidizer is oxygen-supplied as ambient atmospheric air, compressed air, or sometimes as pure O_2 in high-temperature industrial oxy-fuel systems. In combustion, hydrogen does an exothermic redox reaction with oxygen:

$$2H_2 + O_2 \rightarrow 2H_2O, \Delta H = -571.6\,\text{kJ} / \text{mol}$$

This is an exothermic reaction with the only product being water vapor (H_2O); thus, hydrogen is a zero-carbon emission fuel. When we determine the stoichiometric air requirement, we need to take into consideration that air consists of ~21% O_2 by volume. Therefore, we require 1 mole of O_2 to be supplied. This can be done with the following:

$$\text{Air required per mole of } O_2 = {}^{1}\!/_{0.21} = 4.76\,\text{mol air}$$

Therefore, for the stoichiometric combustion of hydrogen,

$$2H_2 + 1O_2\left(\text{from } 4.76\,\text{mol air}\right) \rightarrow 2\,H_2O$$

This means that 1 mole of H_2 requires around 2.38 moles of air and hence the stoichiometric ratio of air–fuel ratio in mole ratio is 2.38:1. However, since the laminar burning velocity of hydrogen (~2.65 m/s in air) is very high and has a large flammability range (4–75% in air), the burner should be constructed in a manner that a flashback should be avoided, that is, expecting the flame to travel in the direction opposite to the flow of the mixture, toward the mixing zone. Premixed burner systems use flame arrestors, porous media, or sonic velocity nozzles to counter this effect.

There are two main types of mixing strategies, which are the following:

Premixed Burners

Those using premixed hydrogen burners mix hydrogen gas and air in controlled proportions before lighting to create a uniform and readily combustible mixture. This allows a more complete and cleaner combustion with lower CO and NO_x emissions when working lean (with excess air). Premixed flames are short, blue, and stable, with high thermal efficiency and an adiabatic temperature greater than 2000°C. However, because hydrogen has a high flame speed of 2.65 m/s, a broad flame flammability range of 4–75% in air, and a low ignition energy of 0.02 mJ, premixed systems tend to flashback, where the flame moves upstream to the mixing zone, leading to explosion and equipment damage. For this reason, burner designs include flame arrestors, high-velocity flow channels, and swirl stabilization mechanisms to secure the flame. These challenges aside, premixed hydrogen burners are found in many industrial applications requiring low-emission/high-efficiency heating, for example, glass manufacturing, metal processing/ hydrogen-powered boilers.

Diffusion Burners

Diffusion hydrogen burners are designed to admit hydrogen and air independently so that they may not combine until the flame zone where the combustion takes place. Such a design greatly minimizes the possibility of flashback, and the system is safer and easier to use than premixed burners. However, since the fuel and oxidizer do not mix in a uniform manner, the combustion may not be complete, and there may well be local areas of high temperature created, particularly in stoichiometric or fuel-rich surroundings. Such hot spots enhance the creation of NO_x via the Zeldovich reaction with respect to temperature, whereby atmospheric nitrogen and oxygen react at high temperatures above 1800°C, forming nitric oxide (NO), which undergoes further oxidation to nitrogen dioxide (NO_2). To reduce the emission of NO_x, techniques that include staged combustion, inert gas dilution, or flameless combustion are used to reduce the peak flame temperature. Moreover, the thermal efficiency may be further improved by pre-heating the combustion air, adding stability to the flame, and raising the enthalpy of reactants; however, at the expense of higher peak temperatures and the added thermal stress experienced by burner materials. This requires the utilization of heat-resistant alloys or ceramics that are hydrogen-compatible to avoid hydrogen embrittlement as well as mechanical deterioration within hot areas of the reaction (Table 6.1).

6.2.1.2 Step 2: Mixing

The mixing of hydrogen (H_2) with either air or pure oxygen (O_2) is an essential step within hydrogen combustion systems, which controls reaction kinetics, flame temperature, formation of pollutants, and stability of systems. The equivalence ratio (Phi, Phi), which is the ratio between the actual fuel to air ratio and the corresponding stoichiometric weight ratio, defines the mixing ratio; $\Phi = 1$ means stoichiometric

TABLE 6.1
Comparison between Premixed Burner and Diffusion Burner

Feature	Premixed Burner	Diffusion Burner
Mixing Location	Before ignition	At the burner tip
Flame Type	Homogeneous, clean, blue flame	Yellow or orange (soot-prone)
NO_x Formation	Low (especially with lean mixture)	Higher (due to hot spots)
Flashback Risk	High (due to fast flame propagation)	Low
Combustion Efficiency	High (complete combustion)	Moderate
Control Requirements	Complex (needs safety systems)	Simple

combustion, and $\Phi < 1$ means lean combustion. Conditions at stoichiometric proportions yield a balanced reaction whose energy released is ~571.6 kJ/mol with maximum adiabatic flame temperatures of 2300–2800°C, respectively, depending on pressure and pre-heat conditions. This maximizes energy output, but at these high temperatures, the Zeldovich mechanism causes thermally produced NO_x to form as the nitrogen and oxygen in the air gas react at high temperatures (above 1800°C):

$$O + N_2 \rightarrow NO + N, N + O_2 \rightarrow NO + O$$

To reduce this, a lean premixed condition ($\Phi < 1$) is commonly used, which reduces flame temperatures and widens flammability limits. The mixing procedure is critical and could be premixed or diffusion. In premixed burners, the hydrogen and air are mixed before ignition, resulting in a more homogeneous and efficient burn, although flame arrestors and high-velocity flows are required to eliminate the possibility of flashback since hydrogen has a low ignition energy (~0.02 mJ) and high flame speed (~2.65 m/s). In diffusion burners, the air and the hydrogen are injected independently in separate burners, which mix and react in the flame front, which is inherently safer, but the mixing may cause the flame to be non-homogeneous, the local temperature to rise, and increased NO_x is produced. In a scientific sense, efficiency and the safety of the combustion process are controlled by the diffusion coefficient of hydrogen (~0.61 cm^2/s in air at STP), and flame speed, as well as thermal and mass transport in the mixing zone.

6.2.1.3 Step 3: Ignition

Ignition in hydrogen burners means the initial ignition of the combustion reaction of H_2 with an oxidizer (usually air or pure oxygen). It needs another source of energy to get over the activation energy barrier for the reaction. The common ignition sources are the electric sparks, pilot flames, or electric resistance igniters that provide the energy needed to reach a local temperature high enough to separate the hydrogen and oxygen molecules into reactive radicals that ignite the chain reaction of combustion.

Hydrogen is extremely reactive and has an MIE of 0.02 millijoules, a lot lower than that of almost all hydrocarbon fuels, for instance. It follows that even small energy inputs, like a static discharge, could ignite hydrogen–air mixtures. Its autoignition temperature is between 500°C and 580°C, and the flame speed is 2.65 m/s, making ignition easy but dangerous if not controlled. Ignition systems must be designed to keep the fuel–air mixture within flammability limits (4–75% H_2 in air) and to ignite only when the mixture is contained and stable. In premixed burners, ignition occurs near the burner tip or in a controlled chamber when the mixture reaches the proper stoichiometry. In diffusion burners, ignition occurs when hydrogen mixes with atmospheric oxygen at the burner outlet. For safety and reliability reasons, many modern hydrogen burners have automated ignition systems with spark plugs, plasma torches, or glow plugs and flame sensors, UV/IR detectors, and thermocouples to check ignition and prevent delayed ignition, flameouts, and other similar issues. Also, interlock systems prevent ignition when hydrogen concentrations are outside flammable limits or airflow is inadequate.

6.2.1.4 Step 4: Combustion

At the burner head, the hydrogen–air mixture burns quickly when ignited externally by a spark or pilot flame, or an electric igniter. The core reaction is a very exothermic redox reaction between H_2 and O_2 at −571.6 kJ/mol. This reacts almost instantaneously if the conditions are right stoichiometric or lean, and gives off water vapor (H_2O) as the primary product. The released energy raises the flame temperature to between 2000°C and 2800°C, depending on air–fuel ratio, pressure, and pre-heating of reactants. If just enough oxygen is available for full oxidation of the hydrogen under stoichiometric conditions, the adiabatic flame temperature can reach its maximum (2300–2500°C in air and much higher in pure oxygen).

A typical hydrogen flame is usually pale blue to almost indistinct in well-ventilated environments because no carbon-based radicals, such as CH or C_2, give off bright visible light in hydrocarbon flames. Hydrogen combustion produces no soot, smoke, or CO_2; the flame is clean and colorless and almost silent, and is not detectable without sensors. This clean combustion is among the environmental benefits of hydrogen as a fuel. For fluid dynamics reasons, combustion produces high-speed gases and pressure gradients that define the flame shape and velocity field. Radiators like OH, H, and O also exist in the combustion zone and help spread the flame front in chain reactions. Flavor speed for such hydrogen–air mixtures is relatively high (2.65 m/s in stoichiometric conditions), so burner geometry should be controlled well to prevent blow-off or flashback. Importantly, although water vapor is the product of choice, secondary reactions can occur at a sufficiently high flame temperature, including nitric oxide formation via the thermal NO_x pathway with atmospheric nitrogen. This explains why combustion temperature and residence time must be controlled in burner design.

6.2.1.5 Step 5: Heat Transfer

Combustion of the gas hydrogen produces strong thermal energy that is mainly composed of high-temperature water vapor that should be efficiently conducted to a heating surface, chamber, or working fluid to do useful work. Depending on the burner design and application, one or more of the traditional mechanisms of heat transfer are involved, namely, conduction, convection, and radiation. In a stoichiometric mixture, the ignition of hydrogen liberates approximately 286 kJ/mol heat, usually at a standard temperature, with adiabatic flame temperatures of more than 2000–2800°C, depending on the flame propagation mode. Much of this heat energy is transported with the main product, superheated steam (H_2O vapor). It is a hot gas that is fed through the combustion chamber, where it heats up a surrounding surface or process fluid.

1. Convection is the most common process in systems where the hot combustion gases move past the metal tubes, transferring heat to a working fluid (usually water, giving steam). The speed at which convective heat transfer occurs is affected by the temperature difference, the velocity of the gases, and the surface area, and adheres to Newton's Law of Cooling:

$$q_{\text{conv}} = hA\left(T_g - T_s\right)$$

where

q_{conv} = Heat transfer rate
h = convective heat transfer coefficient
A = Surface area
T_g = Gas temperature
T_s = Surface temperature

2. The radiative heat transfer plays an important role in high-temperature systems (e.g., industrial furnaces or kilns) where the energy carried by the thermal radiation emitted by the flame and hot gases is transferred to the chamber walls or load. This is stipulated by the Stefan–Boltzmann Law:

$$q_{\text{rad}} = \varepsilon\sigma A\left(T_f^4 - T_s^4\right)$$

where

q_{rad} =Heat transfer rate
ε = Emissivity
σ = Stefan–Boltzmann constant
T_g = Gas temperature
T_s =Surface temperature

3. If the catalytic hydrogen burner is operating at lower temperatures than the catalyst (platinum or palladium surfaces), then heat is transferred mainly by conduction and low-temperature convection to the nearby surfaces or fluids. Such systems are used for space heating, chemical processing, or cleanrooms where open flames are undesirable.

The temperature is just one of many factors that determine the efficient heat transfer; others are gas composition, turbulence, burner geometry, and material properties. As an example, the specific heat capacity and thermal conductivity of water vapor (the by-product of combusting hydrogen) are much greater compared to those of air or CO_2, so that convective heat transfer is increased, although it can also result in condensation and corrosion on cooler surfaces that is not managed properly.

6.2.1.6 Step 6: Exhaust

Hydrogen combustion produces exhaust that is essentially cleaner than hydrocarbon fuels, since the by-product of hydrogen combustion is largely water vapor (H_2O), with no CO_2, carbon monoxide (CO), or soot, or unburned hydrocarbons, emitted.

$$2H_2 + O_2 \rightarrow 2H_2O, \Delta H = -571.6\,\text{kJ/mol}$$

In the primary reaction, a great deal of energy is released, and superheated steam is the major exhaust component. In addition to steam, the exhaust contains inert nitrogen (N_2) from the combustion air and residual oxygen (O_2) if supplied with excess air. Despite this clean-burning character of hydrogen, however, small amounts of NO_x – mostly NO and NO_2 – form at high flame temperatures (typically above

1800°C) via the thermal Zeldovich mechanism, where atmospheric nitrogen reacts with oxygen in the high-temperature combustion zone. The NO_x formation extent depends on flame temperature, gas residence time in the hot zone, and burner design. Modern hydrogen burners employ low peak flame temperatures and ultra-low NO_x emissions, such as lean premixed combustion, flameless oxidation (FLOX)/steam, or flue gas recirculation/staged combustion to reduce peak flame temperatures and NO_x emissions. The high moisture content of the exhaust may also cause condensation downstream and corrosion of metallic parts unless resistant materials or protection coatings are used.

6.3 TYPES OF HYDROGEN BURNERS

Hydrogen burners are available in a variety of forms, each of which is tuned to the special considerations of hydrogen combustion, including the high speed of flame and the propensity to produce NO_x. Other typical variations are partially premixed burners, which mix hydrogen and air shortly before the burn, enhancing flame stability and decreasing NO_x emissions, as in the current design, in which geometric optimizations are performed, as well as humidification to further decrease emissions and increase safety. Multi-fuel industrial burners are designed to accommodate various fuels, including pure hydrogen, through detailed mixing and combustion regimes such as MILD (Moderate or Intense Low-oxygen Dilution) to offer high efficiency and low emissions. With multiple, smaller nozzles, Micromix burners form miniaturized diffusion flames that have reduced risk of flashback and low NO_x emissions, making them appropriate to use in both industry and gas turbine applications. Catalytic hydrogen burners can be made of materials such as platinum, allowing stable, low-temperature combustion, and are particularly important in hydrogen production systems involving reformers. Hydrogen can be mixed with any other fuel (typically ammonia or methane) on dual-fuel burners, and occasionally by dual-swirl or stratified injection, can be used to limit the shape of the flame and reduce problems such as thermoacoustic instability. Non-catalytic low-NO_x burners/ceramic foams are also applicable, providing hydrogen-rich fuel blend flexibility and very low levels of emissions. Each type of burner is customized according to the different requirements that it must operate in, by considering efficiency, safety, and control of emissions in hydrogen combustion systems.

6.3.1 Premixed Hydrogen Burners

The premixed hydrogen burner is being developed for cleaner combustion, but its design is challenging because hydrogen has unique properties, such as high flame speed/low ignition energy, that cause flashback and operational instability. Research shows that hydrogen addition in premixed burners improves combustion efficiency and CO and unburnt hydrocarbon emissions, but increases flashback risk when burner surface temperatures are higher and the fuel mix contains more hydrogen. Flashback of hydrogen burners can occur in two ways: hydrodynamic propagation of flame upstream/autoignition along hot burner walls, determining the dominant mechanism based on burner geometry/wall temperature. Optimal geometric optimization, like nozzle design adjustment or asymmetry in burner slits, may reduce NO_x emissions and improve resistance to flashback, but narrowing slits alone is not efficient and may increase wall-ignition risk. Advised injection strategies like Micromix or partially premixed configurations have been shown to increase stable operating ranges and reduce flashback when compared with fully premixed designs. Such catalytic approaches as platinum coating burner walls may stabilize ultra-lean hydrogen flames and assure complete combustion, but NO_x emissions increase with higher equivalence ratios. Schmidt et al. developed a low NO_x partially premixed hydrogen burner combining numerical simulations with experimental diagnostics for flame stability improvement and emission reduction. The authors performed the first CFD (RANS) simulations optimizing the burner geometry, focusing on mixture generation at the nozzle exit. A flow body (FB) was introduced to mix hydrogen–air in a short, premixed duct to reduce hot spots and flashback risk and improve flame uniformity. From a scientific perspective, better mixing lowers local temperature gradients and thus NO_x formation since thermal NO formation is governed by the Zeldovich mechanism and is dependent on flame temperature. Experimentally, the burner was tested for flame stability and exhaust gases for NO_x quantification at different equivalence ratios, humidification levels, and thermal loads by infrared imaging (Figure 6.1). Results showed that optimized geometry decreased NO_x emissions by 40% at lean conditions, pH 0.85, and raised the lean operational stability limit. Also, combustion air humidification reduced NO_x formation by another 22–28% near stoichiometric conditions due to lower flame temperatures. The combination of geometric optimization, lean premixing, and humidified air led to a scientific route to suppress NO_x emissions while preserving stable hydrogen combustion (Schmidt et al., 2023).

Vance et al. experimented with the flashback in burner-stabilized hydrogen–air premixed flames to derive a reliable correlation for arriving at flashback limits. The authors have undertaken a systematic process to examine the impact of burner geometry (slit width, spacing, plate thickness), operating conditions (equivalence ratio), on the flashback behavior. With a set of detailed laminar simulations including conjugate heat transfer, thermal diffusion, and radiation, they examined the combined effects of flame stretch, preferential diffusion of hydrogen considered by its low Lewis number, and radiant heat exchange between the burner and the flame. Scientifically, they discovered that all those theory-based advanced flashback models on critical velocity gradient cannot fully explain the phenomenon as they do not acknowledge how stretching causes preferential diffusion effects, which result in a massive increase in local flame velocity in lean hydrogen-based mixtures. In order to eliminate this, the authors developed a new Karlovitz number-based correlation that includes flame stretch, as well as local displacement speed, which effectively collapsed flashback data over a broad range of conditions. Their results indicated that aerodynamic effects (velocity gradients, strain rates) are largely the controlling influences of flashback onset, and thermal effects (burner temperature, pre-heating) are secondary. Their incorporation of these physical phenomena produced a predictive model that successfully describes flashback behavior of hydrogen premixed flames and thus provides insider information in the safe designing of hydrogen combustion systems in the industrial and home environments (Vance et al., 2022).

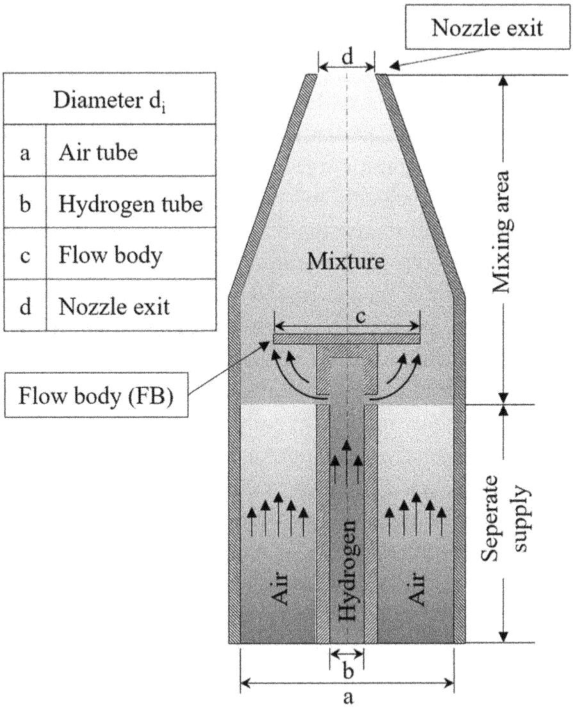

FIGURE 6.1 Schematic sketch of the single nozzle burner with geometric details (Schmidt et al., 2023).

6.3.2 Diffusion Hydrogen Burners

The high reactivity and zero carbon emission of hydrogen make diffusion hydrogen burners a promising technology for clean energy applications. They avoid flashback – a danger with premixed hydrogen flames – and use diffusion-controlled combustion – making these burners safer for industrial use. Research shows that burner design, such as wall material and geometry, influences flame stability, temperature, and emissions: low conductivity materials, such as titanium, may raise flame temperature and energy efficiency but also increase NO_x emissions due to higher peak temperatures. Optimizing burner parameters, including fuel flow velocity and air preheating, may further increase efficiency and control emissions, as higher hydrogen content generally reduces visible flame height and soot formation. By mixing hydrogen with hydrocarbons such as methane or by staged combustion, flame temperatures and NO_x emissions can be reduced while maintaining a balance between efficiency and environmental impact. Because hydrogen is very diffusive, differential diffusion effects drive flame structure and burning rates, especially in turbulent conditions, and can induce localized enrichment and increased burning rates. Microscale and counter-flow diffusion burners for precise control in small-scale applications are also investigated, flame characteristics being sensitive to burner size, gap, and flow rates. Sato et al. studied micro-counter-flow diffusion flames produced by the oxygen combustion of an H_2-CO_2 mixture for understanding microscale hydrogen combustion for small-scale energy devices. They investigated how the burner inner diameter, burner gap, and hydrogen flow rate affect flame stability and character. They measured flame thickness and diameter under controlled conditions with flame image analysis using CH* chemiluminescence. CO_2 was added to hydrogen fuel to reduce flame propagation velocity and make the flame observable. They showed that flame size decreases with increasing burner inner diameter, burner gap, and H_2 flow rate, mainly due to increased flame stretch rates, increased heat loss to burner walls, and radical quenching at the microscales. Then, they introduced a normalized flame thickness parameter inversely related to the square root of flame stretch rate, which gives a universal scaling law for microhydrogen flames independent of burner size and flow condition. Their results also revealed that such findings apply to CH_4-CO_2-O_2 microflames too, showing that the proposed scaling is more general (Sato et al., 2022). Gupta et al. designed a particular kind of diffusion flame burner, hydrogen, which was specifically designed to assist in continuous operation in an electrolytic heavy water reconcentration plant at Trombay in India. The authors acknowledged the shortcomings of the premixed hydrogen oxygen burners, for example, the hazard of backflash, red-hot burner heads, and large pressure drops, hence the use of diffusion-type burners where hydrogen and oxygen are fed separately, with ignition taking place in the flame front due to molecular diffusion. Such a design does not result in the formation of an explosive mixture of gases in the body of the burner, which significantly increases the safety of the work. The authors have developed and tested several burner heads to study the performance at different hydrogen flow rates (2–10 Nm^3/h) and mass velocity ratios (oxygen-to-hydrogen). In a wide range of experiments, they discovered that the flame height is directly proportional to the velocity of hydrogen but starts to stagnate because of the laminar–turbulence transition, and that flame height is inversely proportional to the mass velocity ratio because of the raised mixing degree and the diminished flame zone. Their combustion efficiency was 99.98% and more. The burner was operated safely for more than 1000 h without any flashback or explosion. System pressure drop was low (~19.2 cm water gauge at full flow), quite superior compared with premixed systems. The burner incorporated water cooling, non-return valves, and flame arrestors so that the combustion process would be continuous and stable (Gupta et al., 1964).

6.3.3 Catalytic Hydrogen Burners

Advanced catalytic hydrogen burners employ noble metals such as platinum or palladium as catalysts to burn hydrogen in flameless mode at lower temperatures for safer operation and ultra-low emissions, especially NO_x. These burners are becoming more popular for residential heating, industrial processes, and fuel cell integration because they convert hydrogen's chemical energy to heat with minimal pollutants. Compared with conventional flame combustion, Catalytic Hydrogen Combustion (CHC) produces only water as a by-product and poses fewer safety hazards, such as blow-off and autoignition. Key challenges are the control of catalyst hot spots, prevention of water and sulfur poisoning, and long-term durability, but recently developed catalyst materials and reactor designs provide better performance and stability. Reactor types like fixed-bed, monolithic, and microchannel are optimized for heat and mass transfer, rapid start-up, and stable operation. In some residential applications, CHC systems achieve up to 100% hydrogen conversion rate and 80% heat efficiency. Choice of catalyst support materials and fabrication methods also influences burner performance and life.

In these burners, a combination of hydrogen and air is passed over a catalyst – generally platinum, palladium, or maybe several metal oxides – wherein hydrogen reacts with oxygen at the catalyst surface to give off heat and water vapor but without a flame. This process (CHC) burns hydrogen at lower temperatures than traditional flame combustion, reducing NO_x formation and improving safety, and is suitable for residential heating/industrial applications. The catalyst provides active sites that allow the reaction to occur with very lean hydrogen–air mixtures, sometimes well beyond the normal flammability limits. A burner designed with honeycomb or catalyst-coated structures maximizes contact between the gas and the catalyst for better combustion efficiency and temperature uniformity. Operating parameters like hydrogen-to-air ratio, flow rate, and temperature affect the burner performance, but stable combustion leads to almost 100% hydrogen conversion and high thermal efficiency. Catalytic burners are developed for fast start-up, long-term use with minimal emissions, and are used in everything from water heating units and fuel cells. The main challenges are catalyst durability and resistance to poisoning by water or sulfur, but research is ongoing on these issues for long-term reliability. An industrial-scale CHC system for residential heat supply was proposed by Wang et al. The 1 wt% platinum (Pt)-loaded glass fiber felts were prepared by wet impregnation utilizing Pt's high catalytic activity for low-temperature hydrogen oxidation. In a stainless steel reactor, the burner design employed several layers of these Pt-coated felts for flameless combustion at controlled temperature. Science says catalytic combustion occurs at 500°C lower than flame combustion because the reaction occurs on the catalyst surface, thus reducing thermal NO_x formation and avoiding flashback hazards. The authors experimented with hydrogen conversion, combustion stability, and heat efficiency. They found almost 100% hydrogen conversion under stable operation, and the catalytic reaction was stable with changing the hydrogen flow rate. Furthermore, the system was able to achieve 80% heat efficiency for hot water supply due to heat transfer via a water-based heat exchanger and the high surface area of the Pt catalyst. They also showed that the H_2/air ratio influenced ignition dynamics but not during the stable combustion stage. However, long-term use led to Pt particle agglomeration, performance slightly degraded, and water condensation presented operational challenges at low temperatures (Sarkar & Bhattacharyya, 2012).

6.3.4 Oxy-Hydrogen Burners

Oxy-hydrogen burners, in which hydrogen is burned in air-free oxygen, are under consideration as a high thermal efficiency replacement of classical hydrocarbon burners that can severely lower emissions. These burners have also been found to be rather suitable in industrial and home applications, with higher thermal efficiencies of up to 76% achieved at higher flow rates. Hydrogen burning with oxygen forms nothing but water vapor, so there is zero volume of CO_2 and CO emissions, and with modern designs of the burner, the emissions of NO_x may be maintained far below regulatory set limits. Oxy-hydrogen burners are also stable and safe to use in a wide power range and with renewable hydrogen, decarbonization of fossil-fuel-sandwiched sectors is also achievable. Hydrogen-enriched oxy-fuels promote heat transfer and enhance flame stability with the ability to operate flexibly, even during the extremely dilute conditions needed to generate zero-emission power cycles. Flame stability and turndown are also improved in more advanced burner technologies, including Micromix burner designs, with potential applications to both new gas turbines and combined heat and power stations. Moreover, the oxy-hydrogen burners also do away with problems such as carbon buildup and fuel leakage related to LPG systems, making them very safe and less costly to maintain. Although these technologies have great potential, there is still more work to do on testing and refinement to more widely use them in industries, especially in terms of cost-effectiveness and the ability to combine them with the existing infrastructure. Schwarz et al. studied the combustion properties of hydrogen-enriched natural gas when operating with an oxidizer in oxy-fuel conditions to contribute to the decarbonization of high-temperature industrial processes. The experiment was carried out in a semi-industrial furnace designed with a high-impulse multi-fuel burner designed to run on both something completely hooded (0%) and something totally hydrogen (100%). The experiments were performed at steady-state with fuel input between 100 and 140 kW and thermally loaded conditions representative of the steady-state conditions as simulated by water-cooled lances. To assess the flame temperatures, varying heat fluxes, energy balances, and formation of pollutants, advanced measurement techniques such as gas analyzers, thermocouples, and flow meters were used. The findings indicated that a higher hydrogen content increased the adiabatic flame temperature (to 3080°C maximum), augmented laminar burning velocity and heat transfer to cooling lances, and boosted thermal efficiency (up to 7.8%), respectively. Meanwhile, flue gas temperatures dropped (up to 126 K in the case of pure hydrogen), which testified to improved heat utilization. The addition of hydrogen also suppressed soot formation caused by the high oxidation rate at a higher flame temperature. Although a significant variation in the velocities of the outlets (up to 3.47 times with pure hydrogen) was observed, a stable flame length and overall distribution of temperature validated the flexibility of the burner. According to the scientific view, this increase could be explained by the fact that hydrogen was more reactive and diffused faster, thus promoting more rapid combustion and increasing radiative heat transfer. Lack of carbon substance also reduced the emission of soot and CO_2. The authors concluded that the use of hydrogen instead of natural gas in industrial furnaces did not require significant changes, which is an effective channel that allows energy-intensive industries to reduce the amount of CO_2 emitted into the atmosphere (Schwarz et al., 2024).

Figure 6.2 shows an experimental combustion chamber used to investigate the behavior of a burner in a controlled environment, and it is directly relevant to the working principle of oxy-hydrogen burners. This arrangement provides hydrogen fuel and oxygen via a regulated control line to make sure mixing ratios can be very precise, thus it can operate on pure oxy-fuel conditions. This mixture of hydrogen and oxygen is injected into the insulated chamber by the burner, where the lack of

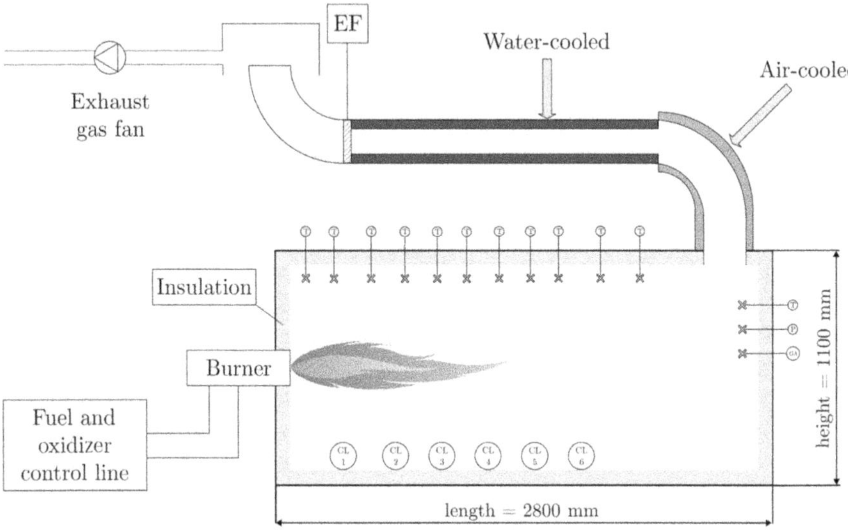

FIGURE 6.2 Sectional view projected through the center of the test furnace, which shows the main part of the cooling and measuring equipment (Schwarz et al., 2024).

nitrogen causes an extremely hot flame (up to ~2800°C) with no heat loss to the inert gas. Flame heat is conducted to the chamber walls, which are fitted with several heat sensors to quantify heat flux at different locations. To control and analyze the distribution of heat, there are water-cooled areas in the chamber that receive a lot of heat and air-cooled areas that aid in maintaining the chamber temperature. An exhaust fan removes the combustion products, which are principally steam (H_2O vapor) and some oxygen, making it continuously operational. This design matches the characteristics of oxy-hydrogen burners, where flames are intense and heat is efficiently transferred, and the formation of NO_x is minimal because nitrogen is not involved in the combustion process. The controlled environment and cooling arrangements in the system also provide safety in handling the highly reactive hydrogen flame during the simulation of heating conditions in an industry. Generally, such a configuration will illustrate how oxy-hydrogen burners can be designed and optimized to have high thermal efficiency, clean combustion, and increased heat recovery, which makes them suitable for applications like metal processing, glass melting, and other high-temperature processes in industries.

6.3.5 Flameless (FLOX) Hydrogen Burners

Flameless (FLOX) hydrogen burners represent an attractive technology for industrial decarbonization and can achieve very low NO_x emissions even with hydrogen, which produces more NO_x than most other fuels due to the high flame temperature. Research shows that flameless hydrogen combustion with proper burner and nozzle design can operate stably and reduce NO_x more than conventional flames with different optimal parameters than natural gas burners. Studies in laboratory and industrial furnaces show that flameless combustion eliminates hot spots and disperses heat more uniformly, resulting in more than 85% NO_x emission reduction, although pure hydrogen and high hydrogen blends still produce more NO_x than methane in some chemical pathways. Computing and experimental work show that even old natural gas burners may be modified for hydrogen or hydrogen-enriched fuels with flameless conditions and low emissions without major hardware changes. Advanced modeling and validation efforts further optimized burner designs, demonstrating that flameless hydrogen combustion can increase furnace efficiency up to 7% over natural gas with improved heat transfer and reduced flue gas losses. Strategies like fuel dilution by CO_2/H_2O and novel injector geometry enlarge the flameless regime toward higher hydrogen concentrations to make the technology open to industrial applications. High-pressure and high-power-density tests confirm the scalability and robustness of FLOX burners for gas turbines and other demanding applications with complete combustion and low pollutant emissions. The influence of nozzle design parameters on exhaust gas characteristics during practical flameless hydrogen combustion was investigated by Kikuchi et al. High NO_x emissions in hydrogen burners were a challenge the authors faced because hydrogen has a high flame temperature. They used an experimental furnace having adjustable burner parameters to study primary and secondary air velocity effects as well as fuel gas flow velocity effects, nozzle's Pitch Circle Diameter (PCD), excess air ratio effects, and pre-heated air temperature. By systematically varying these factors, they observed changes in combustion stability, temperature distribution, and NO_x formation. Scientifically, they found flameless combustion – in which the flame is not clearly visible and the reaction occurs in the whole furnace – when fuel and air are injected separately and slowly, accompanied by exhaust gas recirculation. In this setup, the main drivers of NO_x formation via the thermal Zeldovich mechanism were the diluted oxygen concentration and low peak flame temperatures. Unlike hydrocarbons, stable flameless combustion was achieved even at room temperature without pre-heating for hydrogen. They also found that higher hydrogen fuel velocities > 100 m/s and air velocity ratios > 1.0 Va/Vf were required to suppress NO_x formation. NO_x emissions were reduced over 80% under optimized conditions compared to conventional combustion (Kikuchi et al., 2022).

6.4 DESIGN AND ENGINEERING OF HYDROGEN BURNERS

Hydrogen burner design and engineering require a detailed understanding of the combustion behavior to produce safe, efficient, and environmentally friendly systems. Unlike hydrocarbon fuels such as ethanol, which has a very high laminar flame speed 2.65 m/s, a very wide flammability range (4–75% in air), and very low ignition energy 0.02 mJ, hydrogen is an excellent fuel but a safety challenge as well. All these properties call for burners with precise control of fuel–air mixing, flame stability, and flashback protection against backward travel of flame into the premixing zone. Burner geometry is essential: design features may include swirlers for enhanced turbulence/mixing, turbulence/mixing injectors for even distribution of fuel, and flame arrestors/quench meshes for stopping flame propagation upstream. Depending on the operation need, burners may operate either in premixed mode with fuel and air mixture before ignition for cleaner combustion or diffusion mode with fuel and air at the flame front for safety. To withstand the high combustion temperature of hydrogen (2800°C), burner components are made of high-temperature alloys like Inconel/Hastelloy or ceramics resistant to oxidation and to thermal shock. In addition, hydrogen may embrittle metals, so careful material selection is required for long-term durability. Check valves, pressure regulators, flashback detection systems, and cooling jackets are safety devices built into the engineering process for safe operation under variable load conditions. Emissions minimization – especially NO_x – is another key engineering goal. Hydrogen combustion produces neither CO_2 nor CO, but at high temperatures it can produce thermal NO_x via the Zeldovich mechanism. Engineers attenuate this with lean premixed combustion that lowers flame temperature; steam or flue gas refrigeration to dilute the flame; or FLOX, where combustion is uniform without a flame, thus preventing hot spots. Catalytic burners also permit flameless hydrogen combustion with catalysts such as platinum that allow oxidation at lower temperatures with near-zero NO_x emissions.

Research shows that flameless hydrogen combustion with proper burner and nozzle design can operate stably and reduce NO_x more than conventional flames with different optimal parameters than natural gas burners. Studies in laboratory and industrial furnaces show that flameless combustion suppresses hot spots and disperses heat more uniformly, so more than 85% NO_x emission reduction is achieved, although pure hydrogen or high hydrogen blends still produce more NO_x than methane due to specific chemical pathways. Computing and experimental work show that even old natural gas burners may be modified for hydrogen or hydrogen-enriched fuels with flameless conditions and low emissions without major hardware changes. Modern burner development uses CFD to model flame propagation, mixing, and heat transfer, allowing optimization before constructing physical prototypes. The resulting designs are now in operation in low NO_x hydrogen boilers, industrial gas turbines, glass and metal processing furnaces, and residential heating systems. Furthermore, pure oxygen oxidiser oxy-hydrogen burners with high flame temperatures and NO_x elimination due to the absence of nitrogen are ideal for special industrial applications. Advanced modeling and validation efforts further optimized burner designs, demonstrating that flameless hydrogen combustion can increase furnace efficiency up to 7% over natural gas with improved heat transfer and reduced flue gas losses. Strategies like fuel dilution by CO_2/H_2O and novel injector geometries enlarge the flameless regime toward higher hydrogen concentrations in order to make the technology open to industrial applications. Tests of the scalability and robustness of FLOX burners for gas turbines and other demanding applications consistently show complete combustion with low pollutant emissions.

6.5 APPLICATIONS OF HYDROGEN BURNERS

Hydrogen burners are increasingly being used in numerous industries owing to their prospective in minimizing greenhouse gases and expansion in their energy efficiency. Hydrogen burners are important when it comes to producing clean and efficient heat in industries, power plants, and residential systems, owing to the fact that they produce zero CO_2 emissions. They facilitate the shift toward low-carbon sources of energy in their application in industrial furnaces, boilers, and gas turbines.

Flare systems incorporated in the refineries improve safety, and this is as a result of the hydrogen burners that occur without any possible release of hydrogen. Also, the fact that they are being utilized in waste burning and commercial heating reduces the pollutants, meaning that they are compliant with the environment and would be sustainable.

6.5.1 Industrial Applications

Hydrogen burners in industrial applications are also becoming highly crucial as a carbon emission-free alternative to conventional fuel, with high thermal efficiency. An example is in steelmaking electric arc furnaces (EAFs), where hydrogen burners are under development to replace natural gas burners to provide additional heat when melting steel. Such a replacement would have a substantial decrease in the amount of CO_2 emissions, as the chemical reaction of hydrogen burns only to form water vapor. Previous pilot projects, like the ones established by ArcelorMittal and SSAB, have proved the efficiency of a hydrogen-assisted EAF that does not modify the quality of steel produced. In the tray processing of glass manufacturing, hydrogen burners are commonly used in the melting and forming of glass, especially Oxy-hydrogen burners. The temperatures of their flames exceed the 2800°C threshold and can be used in order to quickly melt and shape without contaminating it with carbon-based fuels. Indicatively, after Saint-Gobain and other glass companies incorporated hydrogen burners in their line, they are capable of manufacturing high-quality glass and minimizing the carbon footprint of the furnace. Hydrogen burners are employed in metal processing, particularly in forging, annealing, and brazing, where a high-intensity, nearly non-oxidizing flame is important to avoid oxidation and discoloration of the surface. As one example, hydrogen-fired forging furnaces provide uniform and clear metal appearance, freeing metal scales and resulting in better products and cutting down costs involved in post-processing. Hydrogen combustion technologies are designed to suit unique applications, such as metallurgy, with a company such as Linde developing that technology. Equally, ceramic kilns are highly impacted using hydrogen combustion because its generation provides even heating and limits overheating and repetitive burning damage that may lead to flaws such as breakage or warping within the produced ceramic products. When there is an application of advanced low-NO_x designs or flameless combustion, NO_x emissions are also reduced due to the use of hydrogen burners. Several Japanese and German ceramic manufacturers have experimented with hydrogen firing in kilns with encouraging results, in addition to reducing CO_2 emissions, pointing to economic advantages as well.

6.5.2 Energy Sector

Hydrogen burners are also becoming a center of technology in the energy sector in delivering low-carbon and sustainable energy sources. Manufactured gas turbines that feature hydrogen-compatible burners use hydrogen-rich fuels to take advantage of the high flame velocity and clean combustion properties of hydrogen. With the help of lean premixed combustion strategies, it is possible to have these turbines reduce thermal NO_x emissions considerably, which is known to occur in a high flame temperature environment, and yet continue to exhibit a high degree of combustion efficiency and stable functioning. Hydrogen fuel does not produce carbon, so it does not produce CO_2 emissions and meets worldwide climate change carbon reduction objectives. Hydrogen burners can also be used in industry and district heating boilers, in which they may also complement the generation of power. Smaller burners are also becoming available to take the place of natural gas systems. Such boilers may also include state-of-the-art burner technologies, like swirl-stabilized nozzles and flame arrestors, to ensure safe and maximum air–fuel mixing takes place. A low-NO_x burner configuration offers an added benefit of increasing heat transfer efficiency and, hence, is best suited to large-scale heat networks. These applications combined will assist the shift to hydrogen-based energy systems, incorporate the renewable hydrogen generated through electrolysis, and deliver cleaner, more versatile heating and power to the future energy environment.

6.5.3 Aeronautical Industry

The aeronautics industry is developing low NO_x hydrogen burners as part of the decarbonization of aviation and the environmental footprint of aircraft propulsion systems. With its high specific energy and zero carbon emissions upon combustion, hydrogen is a good substitute for traditional jet fuels; its high flame temperature and high flammability range, however, hinder thermal NO_x control. To meet these challenges, advanced combustion technologies like Lean Direct Injection (LDI) and Micromix combustion are applied to experimental hydrogen-powered engines. The LDI concept directly injects the hydrogen into the combustor as a lean fuel/air mixture to obtain low flame temperatures and low NO_x production conditions. The flame is kept short, evenly distributed, and stable under different power settings by injecting hydrogen through several small injectors, avoiding NO_x hot spots that normally form. In contrast, the Micromix combustion concept divides the hydrogen supply into many microflames that burn ultra-lean and thus reduce NO_x emissions and flashback risks. They are arranged in microflames over a combustor surface to give uniform heat release and good flame stability at high engine load and altitude. Scientifically, such designs exploit the fast reaction kinetics of hydrogen and high diffusivity for good mixing with air and control local flame temperatures for suppressing NO_x formation via the Zeldovich mechanism. Additionally, they are modular, allowing adaptation to different engine geometries and operation requirements. Combining these burner technologies, the aeronautics industry is developing next-generation hydrogen-powered aircraft engines that meet stringent emission standards, maintain high combustion efficiency, and support sustainable aviation.

6.5.4 Domestic and Commercial Applications

Domestic and commercial uses of hydrogen burners are on the rise because of the move toward cleaner energy solutions by the world community. Hydrogen-compatible burners are under development in these environments to be used in cooking stoves, ovens, and commercial kitchen devices. Hydrogen burners produce no

CO_2 on burning as opposed to conventional natural gas burners, making them environmentally friendly. Also, these burners emit low concentrations of CO and lower concentrations of NO_x when they operate under lean premixed conditions, which allows them to control the flame temperature and the mixture of air and fuel. The fast flame speed and the quick ignition characteristics of hydrogen lead to more efficient fuel combustion in cooking systems so that the desired heat level is attained within a relatively short time and with less energy. High-tech designs are equipped with flame arrestors, multi-port injection systems, and swirl-stabilized flames to combine performance and safety. A wide range of contemporary equipment also allows the use of hydrogen-blended fuels (e.g., blends of natural gas or biogas combined with hydrogen), providing flexibility in fuel use and allowing a smooth transition to hydrogen without the massive infrastructural transformation. In scientific terms, the clean burning of hydrogen in such appliances minimizes the generation of incomplete combustion products and particulates, leading to enhanced indoor air quality, which is essential in commercial kitchens. Also, the carbon-free nature of the fuel negates the deposition of soot, which reduces maintenance costs. Hydrogen-ready household burners have been shown to be feasible in pilot programs in countries including Japan, the UK, and Germany, and this has helped establish that natural gases could be replaced by hydrogen with no disruption to the current system.

6.5.5 Environmental and Safety Applications

Hydrogen may be produced as a byproduct in a refinery or chemical plant in a process like catalytic reforming, steam cracking, or hydro processing. If this hydrogen is permitted to escape without any control, it can be a serious safety concern because of its broad flammability range and its low ignition energy. To reduce such risks, flare systems are installed with the use of diffusion burners made of hydrogen that can burn off the excess gas that did not escape before it could build and overwhelm the flare system to release it into the atmosphere. Such burners work by forcing out hydrogen and ambient air separately, and the termination of combustion could only be held at the flame front, where flashback can hardly happen. Hydrogen has very high variable diffusivity, and so it mixes well with air, thus forming a stable flame and fully oxidizing into steam. Flare systems improve the environment since the carbon dioxide or particulate emissions are reduced because flare systems convert hydrogen into harmless steam, and avoid explosions.

Likewise, hydrogen is used in industrial incinerators where hydrogen burners are used as sources of clean, high-temperature fires needed to break down hazardous organic compounds and other pollutants. A high flame temperature of hydrogen (up to 2800°C) makes thermal destruction of waste material very efficient, and the fact that the fuel does not contain carbon implies that no soot and CO_2 are produced. This brings a cleaner burn with very small residual emissions. Moreover, all that combustion left in the hydrogen is water vapor, which makes the environmental performance of such facilities much better. Scientifically, such applications as the use of hydrogen burners for this purpose assist in controlling the thermal NO_x discharges, associated with inclusions of such practices as staged combustion or FLOX, lowering the maximal temperatures of flames.

6.6 COMBUSTION DYNAMICS OF HYDROGEN

The combustion dynamics of hydrogen differ from those of typical hydrocarbon fuels because hydrogen has fast chemical kinetics, high diffusivity, broad flammability limits, and low ignition energy. They affect how hydrogen flames behave under various operating conditions, including flame stability, emissions, and burner performance. Hydrogen flames are typically thin and compact and burn rapidly while settling near the burner nozzle. This rapid stabilization improves combustion efficiency but introduces flashback risk as the flame travels backward into the premixing zone. Therefore, the flow velocity at the nozzle must be greater than the flame speed, and safety devices such as flame arrestors and swirl stabilization must be considered in the design of burners.

The excess air lowers flame temperature and NO_x emissions in lean, fully premixed hydrogen–air flames, but produces complex interactions with high-frequency acoustic modes in the burner or combustion chamber as well. These interactions often produce isolated, oscillating flames sensitive to flow perturbations compared to methane flames, whose oscillations are usually low-frequency collectives. The low Lewis number of hydrogen (Le 1), where thermal diffusivity is greater than mass diffusivity, favors preferential heat diffusion before the flame front. This increases local burning velocity and flame susceptibility to disturbances, making hydrogen flames more susceptible to thermoacoustic instability. These instabilities are characterized by rapid pressure and heat release oscillations that may cause damaging vibrations in combustion systems such as gas turbines.

Combustion dynamics change further when hydrogen is enriched in hydrocarbon flames. Flare length becomes shortened, the zone becomes compacted, and overall burning velocity increases. This changes the flame topology from a diffuse hydrocarbon shape to an anchored flame cone. Hydrogen enrichment also affects thermoacoustic coupling by reducing the phase delay between unsteady heat release and pressure oscillations that can amplify combustion instabilities. A higher flame temperature for hydrogen-enriched flames also increases the speed of sound in combustion gases, modifying the acoustic resonance characteristics of the chamber and possibly creating new instability modes.

For the micro- and mesoscale combustion systems, hydrogen properties are further enhanced due to the high surface-to-volume ratio and large heat losses to burner walls. Hydrogen is a highly diffusive gas with fast reaction kinetics, so flames can remain stable when hydrocarbon flames would burn out. However, depending on flow velocity and geometric configuration, such as nozzle diameter, burner spacing, and confinement, hydrogen flames may transition from stable steady-state combustion to pulsatory burst dynamics in which the flame repeatedly extinguishes and rekindles. These pulsations result from nonlinear interactions between combustion chemistry, vortex shedding, and acoustic pressure waves. Vortices near the burner exit disrupt the flame front and release oscillatory heat, while pressure waves in confined geometries amplify instability. The combination can cause limit-cycle oscillations, where the flame has periodic bursts of combustion.

Scientifically, these dynamic behaviors are related to hydrogen's thermo-diffusive properties and the interaction of chemical reaction time scales with flow and acoustic time scales. At the microscale, the similar timescales of reaction, vortex formation, and acoustic oscillation give nonlinear flame dynamics. Optimizing flow rates, burner geometry, and heat loss management can suppress pulsatory bursts and maintain stable microflames. This is especially important for the microcombustors, MEMS-based power devices, and portable hydrogen energy systems design, which requires high stability and efficiency.

6.7 EMISSION CHARACTERISTICS AND ENVIRONMENTAL BENEFITS

Emission characteristics are all about the nature and concentration of pollutants emitted by various sources like vehicles, ships, industrial plants, and power plants. Such emissions are usually gases and particles like NO_x, CO, volatile organic compounds (VOCs), and particulate matter (PM2.5 and PM10), some of which are harmful, as well as greenhouse gases like CO_2. All the pollutants have their unique environmental and health effects; for example, NO_x causes the formation of ozone and smog, and PM 2.5 is associated with heart and respiratory illnesses. These emissions vary with fuel type, effective combustion, and emissions control technology. Environmental benefits, on the other hand, are the positive consequences of enacting laws that limit the emission of harmful substances or even create no emissions. Such advantages are associated with better air quality, fewer greenhouse gases, less acid rain, and better health of the population because of fewer pollution-related diseases. For example, there have been demonstrated results on the significant reduction of major pollutants associated with the introduction of vehicle emission control policies, including, and limited to, the adoption of cleaner fuels, the use of catalytic converters, and higher standards concerning vehicle emissions. Another viable example of such an effect is witnessed in the case of Hubei in China, where numerous policy options to address vehicle emissions resulted in great environmental results. The regional research confirmed the effectiveness of these policies, as they decreased the concentrations of NO_x, CO, VOCs, and particulate matter (PM), leading to cleaner air and a visible drop in pollution-related health risks. The study also indicated that environmental and health benefits depended on the form and strictness of the policies adopted. Scientifically, such gains were attributed to the decline in precursors of the atmosphere to the generation of smog, and secondary forms of particulate matter that had a direct impact by decreasing the levels of exposure to pollution.

6.7.1 Industrial Sources

For industrial sources like copper smelting plants, emission characteristics are especially important as these factories produce high-temperature particulate material and heavy metals like arsenic, lead, mercury, and cadmium. To reduce these emissions, plants use pollution control devices like electrostatic precipitators /ESPs /baghouse filters/wet scrubbers to catch and remove fine particles from exhaust gases. Such technologies can have over a 99% decrease in particulate emissions and dramatically lower the concentration of harmful contaminants in the atmosphere. This high efficiency is attained by means of electrostatic attraction, fabric filtration, and liquid scrubbing, which trap particles before they are released from smokestacks. However, despite these dramatic reductions, some heavy metals escape into the environment in forms that are not filtered by conventional devices. It happens because some heavy metals volatilize at elevated temperatures and remain as gaseous compounds in the exhaust or are bound to ultrafine particles (nanometer-scale) that can pass through standard filtration systems. They're more dangerous because they penetrate far into the human respiratory tract and deposit toxic metals in tissues. Also, these emissions may collect on soils and water bodies and remain a long-term environmental problem. The persistence of these unfiltered heavy metal emissions suggests that current pollution control technologies are effective but not sufficient to eliminate all risks. Therefore, comprehensive monitoring programs should monitor not only particulate concentrations but also chemical forms and sizes of emitted metals. Advanced monitoring techniques such as continuous emission monitoring systems (CEMSs) and speciated metal analysis can reveal residual emissions and guide the design of future control technologies.

6.7.2 Transportation Sector

Other factors that greatly determine the emission characteristics in the transportation industry include the type of engine, the properties of fuel, and working conditions. The mixture of these parameters could affect the concentrations of NO_x, and this can lead to pollution of a wide range of issues, not only including but also exceeding CO, VOCs, PM, and greenhouse gases. As an illustration, road transport has promoted the use of blended gasoline containing ethanol as a cleaner fuel compared to the usual gasoline; this is because ethanol is blended with oxygen, which allows for more combustion. It can cause a decline in CO and PM emissions. Nonetheless, research has indicated that an elevated ethanol content could even end up in elevated emissions of acetaldehyde and other oxygenated organic compounds, which have the ability to bring secondary air pollution and smog generation. Therefore, inasmuch as ethanol blends come with some environmental advantages, they can raise new pollutant issues that are to be addressed by regulation.

6.7.3 Marine Sector

In maritime transportation that traditionally uses heavy fuel oil, the use of liquefied natural gas (LNG) as a substitute fuel has been promoted toward the aim of curbing emissions of sulfur oxides (SO_x), NO_x, and PM. It is estimated that LNG combustion results in less particulate and sulfur-containing substances as compared to heavy oil. So, LNG is more environmentally friendly than heavy oil. Nevertheless, studies suggest that LNG engines can also produce more CO and (less frequently) unburned methane (also known as methane slip), which is an especially strong greenhouse gas. These fugitive emissions could serve to negate some of these climate advantages because of the switch to LNG. In addition, the LNG engines show engine-dependent variations to the point of operations, for example, varying engine load and maintenance procedures, which further impact the emission results.

6.7.4 Environmental Benefits

Some environmental benefits make hydrogen combustion a good clean energy solution. This decreases the CO_2 emissions, which contribute to global warming reduction. In contrast to hydrocarbon fuels, combustion of hydrogen produces no soot, CO, or unscathed hydrocarbons, thus producing cleaner exhaust gases and reducing urban air pollution. And the lack of sulfur compounds and CO_2 also prevents acid rain from developing. Hydrogen can also be produced from renewable energy sources through electrolysis with renewable electricity – a carbon-neutral fuel cycle.

Green supply chains are integrated supply chains in which environmental considerations are incorporated at every step of the supply chain, beginning from raw material sourcing to production, distribution, and end-of-life product management. Such systems seek to achieve economic efficiency while preserving environmental sustainability. Supported by government regulations and incentives like tax benefits, subsidies for green technologies, or emission standards, green supply chains can promote industrial growth while protecting the environment. These measures encourage enterprises to use cleaner production processes and energy-efficient and low-emission technologies. However, successful green supply chains depend on strong enforcement mechanisms. Some companies free ride without enforcing sustainable practices themselves, without strict monitoring and penalties. This blocks collective progress because emission reductions achieved by proactive companies may be offset by non-compliant firms. Moreover, inconsistent regulation enforcement between jurisdictions or sectors can create loopholes that discourage the wide adoption of green practices. Coordinated efforts between enterprises are also important, as supply chains usually involve several suppliers/manufacturers/distributors/retailers. Emission reduction must involve these stakeholders sharing technology, best practices, and information on sustainable operations. Clear emission targets, transparent reporting frameworks, and penalty systems for non-compliance could provide strong government oversight to facilitate this collaboration. Mandatory carbon disclosure programs or supply chain audits, for example, have helped ensure that all participants meet environmental standards.

6.8 CHALLENGES AND BARRIERS TO ADOPTION

Several technical and systemic obstacles prevent the widespread use of hydrogen burners despite their adoption. High production cost of hydrogen, especially of green hydrogen, which is much more costly than regular fuels, can be the most important obstacle. This is combined with the high cost of retrofitting or installing new hydrogen-compatible burners with high safety and emission control technologies. Also, the absence of an infrastructure of hydrogen distribution, such as pipelines, storage, and fueling stations, reduces the supply of fuel and also increases operational expenses. Technically, the supply chain of specialized parts, including hydrogen embrittlement resistance material and low-NO_x combustion catalyst, is undeveloped. Industries are faced with more uncertainty in terms of systemic barriers such as complex approval processes, unreliability in regulations, and variable safety standards. In addition to this, current funding and incentive programs are problematic and do not support businesses converting to hydrogen technologies in terms of adequate financial resources. All these are aggravated by a lack of coordinated policies and the development of infrastructure. To make them massively adopted, governments should invest in hydrogen infrastructure, streamline regulatory rules, and increase financial incentives, and industries should keep innovating burner designs to achieve greater heating efficiency and lower costs.

6.8.1 Technical Challenges

The complexity of hydrogen combustion presents some of the key engineering challenges in the introduction of hydrogen burners, as one of the key technical issues associated with hydrogen combustion. The high flame speed of hydrogen and low ignition energy cause the flame to be very reactive, and flashback occurs easily, whereby the flame moves up into the premixing area in the burner. This effect can be extremely hazardous, and special burner designs with flame arrestors and high velocity nozzles, and swirl stabilizations are needed to avoid accidents without sacrificing efficiency when maintaining the combustion. Meanwhile, the high temperatures of combustion of a hydrogen fuel lead to potentially enhanced generation of thermal NO_x by the Zeldovich mechanism, even though hydrogen is a carbon-free fuel. This requires the exercise of advanced low-NO_x technologies, including lean premixed combustion, FLOX, or steam dilution to reduce emissions of pollutants. Also, the hydrogen interaction with the materials creates an extra complexity. Hydrogen may bring about metal embrittlement, diffusing in metals, contributing to weakening their structure, and making the process of high-temperature degradation of burner materials faster. To address these concerns, manufacturers need to use special high-temperature alloy materials or ceramic materials, which considerably increase the cost of manufacturing hydrogen burners. Additionally, in hydrogen flames, the dynamics of combustion are very sensitive to acoustic oscillations, disturbances, and turbulence. Such interactions may cause nuisance oscillations or unstable flame behavior, especially in gas turbines and industrial burners; hence, flame stabilization is an important design issue. Therefore, a compromise between safety, low emissions, the material life, and stability not only calls into action advanced engineering but also prudent optimization of the burner design.

6.8.2 Economic Barriers

A major economic barrier to hydrogen burner adoption is their production, installation, and operation. Most hydrogen currently comes from natural gas (gray hydrogen) in processes such as steam methane reforming that produce CO_2 and violate environmental goals. The cleaner alternative, green hydrogen produced by water electrolysis with renewable energy, is much more expensive because electrolyzers and renewable electricity are expensive, as well as advanced storage technologies. This cost gap limits green hydrogen to small-scale use compared to traditional fossil fuels. Other than fuel costs, industries will require substantial capital investments to adopt hydrogen technology. It is expensive upfront to retrofit existing facilities with hydrogen-compatible burners, storage systems, and safety equipment like flame arrestors, leak detection sensors, and high-pressure pipelines. Those expenses can deter industries – especially those with thin profit margins – from installing hydrogen burners despite the environmental benefits.

6.8.3 Infrastructure Limitations

An additional barrier to the wide deployment of hydrogen burners lies in a lack of robust hydrogen infrastructure for supply, storage, and integration into existing energy systems. And unlike natural gas with its vast pipeline and distribution network, hydrogen infrastructure is in its infancy. Deficiency of large pipelines, storage facilities, and refueling stations prevents a steady and cost-effective supply of hydrogen for industrial users that require large volumes of fuel. This constraint delays the scale-up of hydrogen burner technology in both industrial and commercial fields. Storage issues compound the problem. Hydrogen is the lightest, smallest molecule and thus requires high-pressure storage at 350–700 bar or cryogenic storage at very low temperatures (253°C) for practical energy densities. Neither method requires specialized tanks/compressors/insulation systems which increase costs and technical difficulty of industrial operations. In addition to this, the easily leaking nature of hydrogen and the invisible flame make storage systems extremely safety sensitive and require very specific handling and monitoring protocols.

Beyond supply and storage issues, retrofitting constraints are another major obstacle. Existing burner systems, especially those designed for natural gas, are not intrinsically compatible with hydrogen due to combustion characteristics such as flame speed, ignition properties, and heat release profiles that differ between the two gases. To use hydrogen safely and effectively, these burners need new burner nozzles, advanced control systems, flame arrestors, and upgraded safety devices. These upgrades require significant time, cost, and engineering expertise that many industries find prohibitive.

6.8.4 Safety Concerns

One of the key barriers to the adoption of hydrogen burners is how to manage the safety concerns regarding the physical and combustion characteristics of hydrogen. Hydrogen is the lightest and smallest molecule and therefore has a high propensity to leak through sealings, joints, and microcracks in pipes and storage tanks. Any small leaks may accumulate in an enclosed or poorly ventilated area and cause an explosion risk since hydrogen has a broad flammability range (4–75% in air) and a low ignition energy. These qualities make leak-tight systems much more important and technically challenging than is the case with traditional fuels. To make matters worse, it is hard to notice a leak of hydrogen and flames. When burning, hydrogen flames have a barely visible pale-blue flame that cannot be easily seen in daylight, which requires specific optical or infrared detection instruments. They have minimal radiant heat compared with the flames of hydrocarbons, making early detection even more difficult. This will require installation of avant-garde software for flame detectors, leak detection systems, and continual monitoring devices, all of which increase the cost of operations and maintenance. Moreover, the safety regulations and requirements associated with the use of hydrogen burners are also rather strict. These codes stipulate strict safety measures such as explosion-proof parts, well-built ventilating systems, and emergency shutdowns. Although these measures are effective in providing safe operation, they contribute to the regulation and make both the installation of the burner and its operation very expensive.

6.8.5 Regulatory and Market Barriers

Regulatory and market issues of lack of standardization, unclear policy support, and market readiness also act as a barrier to the adoption of hydrogen burners. There is no presence of uniform international codes and benchmarks currently existing that direct hydrogen burners, their infrastructure, and the safety measures taken. This slow harmonization causes confusion amongst the manufacturers and end users, since they have to contend with various regional requirements and approval procedures, which delay their commercialization processes, and also add more expenses in their development process.

Moreover, the policy can be crucial in promoting the uptake of hydrogen, but there are mixed or ambiguous regulatory frameworks in most jurisdictions. Implementation of hydrogen technologies, such as burners, is quite reliant on government incentives, subsidies, and definitive directions toward emission reduction. An industrial sector will not take large risks of investment, where policy is piecemeal or not long-term (or there are fears of policy changing in the future or financial incentives being withdrawn). Such ambiguity negates the confidence in the hydrogen market, still, as a result, slows innovation and the advancement of infrastructure. Moreover, most of the industries would be conservative enough to adopt new technology without economic merit and a regulatory framework in the market. When the fuel price of hydrogen is very high, the infrastructure is not well developed, and there is no guarantee of market stability in the future, businesses will just not want to invest in their hydrogen systems. Surprisingly, there are no major demonstration instances and favorable commercial case studies that can actually support this conservative attitude.

6.9 CURRENT STATE OF IMPLEMENTATION AND PROGRESS IN USE OF HYDROGEN BURNERS

Hydrogen burner use has advanced significantly over the past years, as the rest of the world attempts to decarbonize its industrial operations toward adopting cleaner energy vectors. Hydrogen burners are in trials and application today in heat-intensive industries, power generation, and in specific applications are delivering encouraging results on effectiveness and emission minimization. Their integration is, however, restricted to pilot programs and niche applications, as there are economic and infrastructural barriers to integration.

6.9.1 Case Study 1: ArcelorMittal

ArcelorMittal recently tested a hydrogen-fired burner as part of a programme to decarbonize steelmaking in direct reduced iron (DRI) production. Recent computational modeling of industrial-scale bottom-fired DRI reformers with ArcelorMittal plant data shows that hydrogen and carbon monoxide produced in reformers can reduce iron ore CO_2 emissions in comparison with traditional blast furnaces. In these reformers, the hydrogen-fired burner provides the necessary high temperature for endothermic reactions to enable process efficiency and product quality to be maintained while transitioning to low-carbon steel production (Parvathaneni & Andrade, 2024). More generally, the conversion of natural gas to hydrogen in auxiliary burners, such as those of EAF, has been shown to further reduce emissions but requires adjusting operating conditions because of the different combustion properties of hydrogen (Schüttensack et al., 2024). Such tests at facilities in Germany and France showed that hydrogen combustion could provide enough heat for steel melting and cut CO_2 emissions by almost 90%.

6.9.2 Case Study 2: Saint-Gobain

Saint-Gobain and other major glass manufacturers are using oxy-hydrogen burner technology to decarbonize glass production while enhancing product quality and process performance. Most of the glass melting furnaces use natural gas-air combustion that produces high CO_2 and nitrogen from air as NO_x sources. In contrast, oxy-hydrogen burning uses hydrogen as fuel and pure oxygen as oxidizer, leaving nitrogen out of combustion and reducing CO_2 and NO_x emissions. This transition reflects industry sustainability goals and regulatory pressures to cut greenhouse gas emissions. In trials and pilot implementations, oxy-hydrogen burners have shown up to 60% CO_2 emission reduction, mainly due to the water vapor product of hydrogen combustion. Also, pure oxygen avoids heat losses due to nitrogen dilution, allowing higher flame temperatures (>2800°C) and better heat transfer to the melted glass. It enables fast melting rates, better glass homogeneity, and less defect formation, resulting in a better quality product.

Such environmental gains are at a price. Hydrogen has lower energy content per unit volume and larger flame velocity – flashback takes place in burner designs that can be circumvented by changes in burner design. Moreover, the production of oxy-hydrogen glass is likely to add an extra load of energy by approximately 25%. This is augmented by two factors: (1) the energy intensity of producing hydrogen, specifically when made as green hydrogen through the electrolysis process, and (2) the energy needed to break down and provide oxygen to be burnt. As such, oxy-hydrogen lowers the emissions of CO_2, and is physically and economically practical, mostly alongside the availability of inexpensive renewable hydrogen and cost-effective oxygen-yielding technologies. Nevertheless, the oxy-hydrogen burner companies, like Saint-Gobain, regard the burners as the way toward carbon-neutral glass manufacturing. Renewable hydrogen is becoming less expensive, and oxygen-generating technologies are getting better than oxy-hydrogen combustion. Traditional methods, like natural gas–fired combustion furnaces, used for heating in glass production will be soon replaced by oxy-hydrogen combustion (Gärtner et al., 2021).

6.9.3 More Case Studies

The government-funded Hy4Heat Program in the UK has been a step in the right direction to prove the efficacy of using hydrogen burners in home heating and cooking. This project trialed hydrogen-enabled domestic appliances, such as hydrogen boilers and hydrogen cooking stoves, and demonstrated that it can safely introduce the use of hydrogen into domestic energy systems with relatively minor adaptations to the existing natural gas infrastructure. The significance of special sensors in leak sensing and flame visualization methods was also highlighted in the program to address the safety concerns related to the invisible nature of the hydrogen flame. The hydrogen burners can be an alternative source of household heating and cooking, as the successful trials established that they were safe and practical, ahead of hydrogen-ready houses. Simultaneously, Mitsubishi Power in Japan has advanced considerably relating to the use of low-NO_x hydrogen burners regarding gas turbines utilized in producing power. The company has already been able to demonstrate that stationary turbines can run on fuel mixtures of up to 100% hydrogen, with stable combustion and very low output of greenhouse gases. To counter the problem of NO_x emission that usually accompanies hydrogen due to the intense flame temperature that hydrogen has, Mitsubishi has also implemented superior burner techniques like lean premixing and Micromix burner designs that have helped to reduce the amount of NO_x emission generated. These tests confirmed the opportunity of hydrogen to decarbonize the power industry by allowing carbon-free production of electricity, without compromising performance.

On the same note, the German Aerospace Center (DLR) has gone further and improved the Micromix hydrogen combustion technology in aviation. This is an innovative idea that spreads hydrogen with a widespread of microflames, meaning that flashback is a possibility that is not encountered when using this idea; a common safety hazard in hydrogen combustion. Additionally, NO_x emissions are considerably reduced. It was shown through experimental tests that Micromix burners are very promising burners in the second generation of Hydrogen aircraft engines because of their ability to be stable and burn with low emissions in Hydrogen over a range of different operating conditions. This technology will help the aviation industry adopt zero-carbon emissions with safe and sustainable propulsion sources.

The combination of these case studies by the UK, Japan, and Germany demonstrates how hydrogen burners are being effectively developed to be used in the residential, industrial power generation, and aviation sectors. The underlying lesson of all these projects is that, under the fastidious designs of the burners and suitable safety precautions, sensible combustion controls, hydrogen burning can become highly productive, low-emission, and safe, thus becoming the backbone of clean energy technologies of the future.

6.10 CONCLUSIONS

Hydrogen burners are the confluence of clean energy research and industrial practi cability. They have the perceived benefit of being clean and having excellent burning efficiency, unlike traditional fossil-fuel burners that emit carbon by-products and have low combustion efficiency, consistent with the global decarbonization methods. As shown in this chapter, the successful application of hydrogen burners is not only determined by the characteristics of the fuel but also incorporation of superior designs, improved mixing methods, and superior NO_x mitigation strategies. Although barriers still exist, both technical and economic ones, the hurdles are being lowered through continuous research, pilot-scale demonstrations, and favorable policies.

The illustrations of their use in steel production, glass industries, air transport, and even residential heating point to the obvious step toward further usage. Finally, hydrogen burners are not a mere supplement to conventional combustion systems; they have the potential to change heat-intensive industries, speed up the shift toward renewable energy, and develop a cleaner and more resilient energy world.

REFERENCES

Ayed, A. H., Kusterer, K., Funke, H. H. W., Keinz, J., & Bohn, D. (2017). CFD based exploration of the dry-low-NOx hydrogen micromix combustion technology at increased energy densities. *Propulsion and Power Research*, 6(1), 15–24. https://doi.org/10.1016/j.jppr.2017.01.005

Gärtner, S., Rank, D., Heberl, M., Gaderer, M., Dawoud, B., Haumer, A., & Sterner, M. (2021). Simulation and techno-economic analysis of a power-to-hydrogen process for oxyfuel glass melting. *Energies*. https://doi.org/10.3390/en14248603

Gupta, D., Deshpande, P. G., & Bashyam, S. N. (1964). Diffusion flame burner for continuous combustion of hydrogen. *Industrial and Engineering Chemistry Process Design and Development*, 3(1), 70–73. https://doi.org/10.1021/i260009a016

Kikuchi, K., Motegi, T., Hori, T., & Akamatsu, F. (2022). Influence of nozzle design parameters on exhaust gas characteristics in practical-scale flameless hydrogen combustion. *International Journal of Hydrogen Energy*, 47(49), 21287–21297. https://doi.org/10.1016/j.ijhydene.2022.04.230

Parvathaneni, S., & Andrade, M. W. (2024). CFD modeling of the industrial-scale bottom-fired direct reduced iron reforming process. *Industrial & Engineering Chemistry Research*, 27(63), 11881–11891. https://doi.org/10.1021/acs.iecr.4c01775

Sarkar, J., & Bhattacharyya, S. (2012). Application of graphene and graphene-based materials in clean energy-related devices Minghui. *Archives of Thermodynamics*, 33(4), 23–40. https://doi.org/10.1002/er.1598

Sato, D., Nakachi, S., Honda, K., Obu, K., Katsumi, T., & Kadowaki, S. (2022). Experimental analysis on micro diffusion flames formed by oxygen combustion of H_2–CO_2 mixture using counterflow burners. *Journal of Thermal Science and Technology*, 17(1), 1–12. https://doi.org/10.1299/jtst.22-00012

Schmidt, N., Müller, M., Preuster, P., Zigan, L., Wasserscheid, P., & Will, S. (2023). Development and characterization of a low-NOx partially premixed hydrogen burner using numerical simulation and flame diagnostics. *International Journal of Hydrogen Energy*, 48(41), 15709–15721. https://doi.org/10.1016/j.ijhydene.2023.01.012

Schüttensack, L., Reinicke, A., & Echterhof, T. (2024). Application of hydrogen operated burners in the electric arc furnace. *IOP Conference Series: Materials Science and Engineering*, 9(13). https://doi.org/10.1088/1757-899X/1309/1/012011

Schwarz, S., Daurer, G., Gaber, C., Demuth, M., Prieler, R., & Hochenauer, C. (2024). Experimental investigation of the combustion characteristics in oxy-fuel combustion of hydrogen-enriched natural gas on a semi-industrial scale. *International Journal of Hydrogen Energy*, 49, 323–337. https://doi.org/10.1016/j.ijhydene.2023.07.268

Vance, F. H., de Goey, L. P. H., & van Oijen, J. A. (2022). Development of a flashback correlation for burner-stabilized hydrogen-air premixed flames. *Combustion and Flame*, 243(2022), 112045. https://doi.org/10.1016/j.combustflame.2022.112045

Index

Pages in *italics* refer to figures and pages in **bold** refer to tables.

L

M

N

O

P

S

T

U

V

W

For Product Safety Concerns and Information please contact our EU
representative GPSR@taylorandfrancis.com
Taylor & Francis Verlag GmbH, Kaufingerstraße 24, 80331 München, Germany

www.ingramcontent.com/pod-product-compliance
Ingram Content Group UK Ltd.
Pitfield, Milton Keynes, MK11 3LW, UK
UKHW021507240926
475702UK00035B/611

9781041106272